Be a Successful Remodeling Contractor

R. Dodge Woodson

McGraw-Hill

New York Chicago San Francisco Lisbon London Madrid
Mexico City Milan New Delhi San Juan Seoul
Singapore Sydney Toronto

The McGraw·Hill Companies

Cataloging-in-Publication Data is on file with the Library of Congress

1 2 3 4 5 6 7 8 9 0 DOC/DOC 0 1 0 9 8 7 6 5

ISBN 0-07-144382-7

The sponsoring editor for this book was Larry S. Hager and the production supervisor was Pamela A. Pelton. It was set in Cremona by Lone Wolf Enterprises, Ltd. The art director for the cover was Anthony Landi.

Printed and bound by RR Donnelley.

This book is printed on recycled, acid-free paper containing a minimum of 50% recycled, de-inked fiber.

I dedicate this book to the most important people in my life,
who are: Adam, Afton, Victoria, Jon, and Nate.

Contents

Introduction

Are you tired of pounding nails? Has your boss been giving you a rough time and very little money? Change it. Go into business for yourself as a remodeling contractor. Make some serious money without climbing on the roof. Are you an organized person who can perform as a business owner? If so, you don't need carpentry experience to make a solid income as a general contractor.

I have built as many as 60 single-family homes per year. My background includes remodeling, plumbing, land development, and real estate, but I have lived in and around the trades for about 30 years. This book will show you how to leave your job or step up to a higher level in your business interest to make some substantial income as a general contractor and remodeler. If you happen to be a carpenter, you have an added advantage. But, you don't need any trade experience to be a successful remodeling contractor.

The key to making money as a remodeler is organization. Other elements come into play, and they include the following:

- Field experience is very helpful.
- A good credit rating helps.
- Being able to take control of difficult situations is essential.
- Time management skills will add to your success.
- Having money in the bank never hurts.

All of the items I have just mentioned are major factors, but anyone can do this. You don't have to be a carpenter to be a remodeler. I am not one, and I've been doing this since 1979! You have to be able to run a business and deal with subcontractors if you want to make this career work. It's not that hard for dedicated people.

What's in it for you? Most general contractors add approximately 20 percent to their cost for the cost of a job. You do the math for your area. In general, a $20,000 job relates to a potential remodeler profit of $4,000, or more. This is not a bad return for 7-14 days of work by others if you have the right crews.

The job is not easy. If you are not willing to do a lot of phone work at night, you might want to consider other options. Being a remodeler is stressful, but at the same time it is very rewarding. You can ride around town and say, "I remodeled that." Venturing into remodeling can take two main directions. You can be the person who is making the job come together as a hands-on remodeler, or you can take the white-collar approach. Both work, and the money is not bad.

I've done this a long time and am sharing with you my life experiences, which should save you a lot of trouble and lost money. Learn from my mistakes. That is what this book is about. I want to see my experience help you succeed as a reputable remodeler. Take some time to check out the table of contents and to thumb through the book. The forms alone are nearly priceless for a rookie. Really, this is your ticket to success by learning from my mistakes. I hope you enjoy the read.

C H A P T E R O N E

Going From Carpenter to Remodeling Contractor

You probably bought this book because you are interested in venturing into the construction and remodeling business. And what a business it is! The total value of all private construction work in this country in the year 2003 was about $934 billion, with the residential side of the industry accounting for nearly $470 billion.

Lots of people know that plumbers, electricians, carpenters, painters, and other trades people work evenings and weekends to make extra cash. Many of the trades people I've known, myself included, have moonlighted at one time or another as a way of easing into a full-time business. Can a remodeling contractor get started by working nights and weekends? You bet, and this chapter will show you how.

PRO POINTER

There are lots of ways to enter the construction field: jumping right in and becoming a sole proprietor; joining up with some skilled trade associates to form a small company; or working with an established company while trying to pick up jobs on your own by moonlighting.

Remodelers originate from all walks of life. Many of them start out as carpenters, doing repair work or small remodeling jobs with an eye to becoming a full-scale contractor. I started out as a plumber and grew into remodeling. From there, I went on to build as many as 60 homes a year.

I've met builders who decided that their present occupation as a lawyer, real-estate broker, farmer, firefighter, or policeman was not for them, and so they left secure jobs for a shot at a new and exciting career. Some builders tire of new construction and turn to remodeling for a change of pace. So, most anyone can transition from whatever they are currently doing to remodeling. Making the transition is easier for some than it is for others.

If your background is in construction, you have an obvious advantage over someone who has never set foot on a construction site. While you may have never built a house or remodeled a bathroom on your own, working around other trades gives a good idea of what goes on during the construction and remodeling process. So, how can you leave what you are doing for a living now and enjoy being a remodeling contractor?

PRO POINTER

Field experience alone isn't enough to make someone a good remodeler. It sure helps, but there is a business side to remodeling that also needs to be learned.

Getting started as a remodeler is not easy. Besides a small bankroll and some prior experience in construction, it will be very helpful if you have had some contacts with electrical, plumbing, and other subcontractors and equipment suppliers. And it would also be very helpful to have a relationship with a local bank, even if only to have a checking or savings account there. I'm sure that there are people with the financial resources to start a contracting business in a first-class manner. I never enjoyed this luxury. I had to start at the bottom and crawl up the mountain. At times it seemed as if the mountain was made of gravel, because every time I would near the top, I would slide back down the hill. But I persevered and made it. I think you can, too.

Basic Needs

What are the basic needs for becoming a part-time remodeler? They are less than you might imagine. There are two different types of remodelers. One is the full-service general contractor-a person who hires all the tradesmen required for the job and uses few if any subcontractors. The other type is referred to as a "broker"—a general contractor who subcontracts all or most of the work to other "specialty" contractors (subcontractors).

If you operate as a "full service" contractor, you will have to hire your own tradesmen such as carpenters, electricians, plumbers, framers, and roofers. This means having to meet large weekly payrolls in addition to finding qualified workers and enticing them to work for you.

ARE YOU READY TO BE YOUR OWN CONTRACTOR?

1.	Rate your ability to supervise your project during the day.	0	5	10
2.	Do you have a full time job?	0	5	10
3.	Do you enjoy working with people?	0	5	10
4.	Do you have strong leadership ability?	0	5	10
5.	Are you comfortable around strangers?	0	5	10
6.	How often do you believe what you are told?	0	5	10
7.	Do you act on impulse without thought?	0	5	10
8.	Are you allergic to dust?	0	5	10
9.	Do loud, repetitive noises bother you?	0	5	10
10.	Does your regular job require you to manage people?	0	5	10
11.	Do you enjoy talking on the phone?	0	5	10
12.	How willing are you to work nights, scheduling subs?	0	5	10
13.	How easily are you intimidated by people?	0	5	10
14.	Do you have a shy personality?	0	5	10
15.	Can you make confident decisions?	0	5	10
16.	How much will you research remodeling principles?	0	5	10
17.	Are you sensitive to fumes and odors?	0	5	10
18.	Are you good with numbers?	0	5	10
19.	Do you have a creative mind?	0	5	10
20.	Can you visualize items from a written description?	0	5	10
21.	Do you have strong self discipline?	0	5	10

(continues)

FIGURE 1.1 Are you ready to be your own contractor?

ARE YOU READY TO BE YOUR OWN CONTRACTOR? (continued)

22.	Do you fluster easily?	0	5	10
23.	Do problems cause you extreme stress?	0	5	10
24.	Rate your organizational skills.	0	5	10
25.	Are you vulnerable to sales pitches?	0	5	10
26.	Do you have time to find subcontractors?	0	5	10
27.	Do you enjoy negotiating for the best price?	0	5	10
28.	Is your checkbook balanced today?	0	5	10
29.	Do you utilize a household budget?	0	5	10
30.	Do you feel qualified to control irate subcontractors?	0	5	10
31.	Do you have strong self confidence?	0	5	10
32.	Do you lose your temper easily?	0	5	10
33.	Can you react quickly to unexpected events?	0	5	10
34.	Can you make personal calls from work?	0	5	10
35.	Do you buy bargains, even when you don't need the items?	0	5	10
36.	Is your time financially valuable?	0	5	10
37.	Will you be available to meet code enforcement inspectors?	0	5	10
38.	Do you have a gambler's personality?	0	5	10
39.	Can you be assertive?	0	5	10
40.	Do you enjoy reading technical reports and articles?	0	5	10
41.	Do you retain information you read?	0	5	10
42.	Do you pay attention to small details?	0	5	10
43.	Do you know people who work in the trades?	0	5	10

(continues)

ARE YOU READY TO BE YOUR OWN CONTRACTOR? (continued)

44. Do you trust your judgement? 0 5 10
45. Can you keep accurate, written records? 0 5 10
46. Are you able to do more than one task at a time? 0 5 10
47. How well can you prioritize your day and your duties? 0 5 10
48. Do you feel qualified to coordinate your project? 0 5 10
49. Can you stand to watch your house being torn apart? 0 5 10
50. Are you capable of staying out of the way of the workers? 0 5 10

Add your total score and compare it to the ranges given below to get an idea of your ability to act as the general contractor.

SCORES AND OPTIONS

If your score is 186 or less, seriously consider hiring a professional general contractor. Your answers indicate a weakness to perform the functions of a general contractor. This score may mean you do not have the right personality for the job. Technical points can be learned, but personalities are hard to change. You may be able to accomplish the task if you do extensive research and address your weak points. Keep your quiz answers in mind as you read this book. The book will help you to clearly identify the areas you need to address. For homeowners in this scoring range, hiring a professional is the safest route to take. Before trying to coordinate your own job, read this book and evaluate what you learn. Chances are, you will decide to hire a professional to manage your job. There is nothing wrong with this. Not all people are designed to run construction crews and jobs.

(continues)

ARE YOU READY TO BE YOUR OWN CONTRACTOR? (continued)

If your score is between 186 and 280, you have the ability to learn how to get the job done. Most of the areas you need to work on are remodeling related and can be learned. In this mid-range, you should be able to read enough to attempt the job at hand. Your score indicates some areas of weakness. As you complete this book, note the areas of weakness in your knowledge. Spend the time needed to strengthen these areas. With enough preliminary planning, you should be able to run your own job.

If you scored between 280 and 375, you are a natural. With the right research, you can be an excellent general contractor. The higher your score, the better qualified you are. If you scored near 375, all you will need to do is polish your knowledge of the trades; you already possess the basic qualities of a good general contractor. Even with a high score, you still have a lot to learn. Complete this book and, when you feel completely comfortable with your abilities, move ahead. You will be ready to command your construction crews and save money.

If you operate as a "broker"-type general contractor, you don't need much in the way of trucks, tools, and equipment. The subcontractors you hire will provide their own needs. All you have to do is schedule and supervise their work. And since these subcontractors will generally bill you monthly, your cash flow will occur monthly instead of weekly. However, neither approach is as simple as it seems.

Since you will be working your day job when you get started, you will need an answering service or an answering machine to receive your phone messages. A license to contract work may be required in your region, and a business license will normally be required. You can work from home and meet your prospective customers in their homes. You should invest in liability insurance. Your insurance agent can walk you through various types of policies. And there will be a need to advertise. On the whole, the financial requirements for becoming a part-time remodeler are minimal.

PRO POINTER

It is best for remodelers to have reserve capital to get past unanticipated financial problems, but if you're diligent in your work and if you're a little lucky, you can get by with very little cash.

If you bid jobs accurately and profitably, complete them on schedule, and check your books to ensure that your customers pay you on time, there may not be a need for a large reserve of cash. After all, you will have your regular employment to pay your routine bills and the small additional overhead costs of your building operation.

First Hurdles

There are two hurdles that need to be addressed as you begin to consider a career in construction: any requirements for licensing in the state in which you plan to operate and the lack of a track record and references.

PRO POINTER

It's a good idea to check with your state government to determine if you need to apply for a license and, if so, what the qualifications are. They can vary considerably from state to state.

Let's discuss the licensing issue. Many states require contractors to be licensed before they can operate their business. For example, in the State of Maryland, a license is required before you can work on home improvement or remodeling projects. Before you can get an application, you must pass an exam, and to qualify for the exam, you must pass certain work and financial requirements. In Alabama, you just need four references, proof of insurance, and a net worth of $10,000. In Illinois, most construction contractors, except for roofing contractors, don't need to be licensed.

The second hurdle is the lack of a track record and references. Your customers will probably want the names of references, and they may even want to see examples of your work. When you are starting out, you can't provide references or work samples. This can be a difficult obstacle to clear, but there are some ways to work around the problem.

To overcome the problem of not having references when I started out, I changed the types of advertisements I was running. The ads offered people a chance to have a

job done at a reduced cost if they would allow me to use their names as references. This worked.

Knowledge

How much knowledge of construction do you need to succeed as a remodeler? The more you have, the better off you are. But you can get started with a basic level of knowledge and earn while you learn. It is obviously much easier to supervise people when you understand what they are doing and how the work should be done. But you don't have to be a drywall finisher to supervise drywall work. If a job looks good, you know it. When one looks bad, you can see it.

Code-enforcement officers from the local Building Department will be checking the work at various times, such as during foundation work, framing, and electrical and plumbing rough-ins, to ensure that the work meets code. By sticking close to these officials during their routine inspections, you can pick up a lot of technical information.

PRO POINTER

As a general contractor, you don't have to do any of the physical work involved with construction. Your primary function, once a job is underway, is to schedule and supervise workers.

In theory, you don't have to know much about construction to be a contractor who subs all the work out to independent contractors, but as the general contractor, it is you who will ultimately be responsible for the integrity of the work. Remodelers with the most knowledge of construction are generally much more successful than people who don't know about the home-building process.

If you have good organizational skills and manage people, budgets, and schedules well, you should be able to become a viable contractor. A lot of information can be obtained from watching various subcontractors perform their work. Ask questions, and you'll find subcontractors willing to discuss their work and answer your questions. Reading books and trade magazines will add to your knowledge. Many specialized trade magazines are free, and when you visit a subcontractor's office and see some of those magazines on the table, look for the tear-out forms that will allow you to apply for a subscription. There are numerous books and videos available for do-it-yourselfers that give step-by-step instructions for everything from plumbing to tile work.

PRO POINTER

You can educate yourself by reading and watching tradesmen at work, but field experience is surely helpful.

The Dangers

Even experienced carpenters often don't know how to work up prices for complete jobs. They are not accustomed to figuring in the cost of septic systems, sewer taps, floor coverings, or finish grading. How can you get the best estimates possible when you've never done one before?

Take a set of blueprints to your supplier of building materials. Ask the manager to have someone assess your material needs and price them. Many suppliers will provide this service free of charge, but some won't. Circulate copies of the blueprints to every subcontractor that you will need. It is a good idea to get quotes from more than one subcontractor. Try to select two subs for each trade so that you set up a competitive situation. Have the subs give you prices for all the work they will be expected to do. While your subs and suppliers are working up their prices, you can start doing some homework of your own.

PRO POINTER

One of the biggest dangers for rookie contractors is their lack of experience in pricing jobs.

Take a set of blueprints to a reputable real-estate appraiser. Ask the appraiser to work up either an opinion of value or a full-blown appraisal. This will cost you some money, but it will be well worth it if your remodeling project is a large one.

You can consult some pricing guides to determine what various phases of work will cost. These guides are available in most bookstores, and they have multiplication factors that allow you to adjust the prices to coincide with those in your particular region.

When you get your prices back from suppliers and subcontractors, you can compare them with the numbers you came up with from the pricing guides. You can also look at the difference between the bid prices of your subs and suppliers and the finished appraisal figure. The spread between the bids and the market value represents your potential profit. It should normally relate to about a 10- to 20-percent gross profit. We are going to talk more about pricing and estimating later in the book, but the procedures we have just discussed are the basics.

PRO POINTER

A 15-percent profit might be an average, but the amount varies with economic conditions and the quotes you received from subcontractors and suppliers.

The Internet provides another source for estimating services. The Marshall & Swift website is just one of those sources. There are several cost guides available such as Sweets Repair and Remodel Cost Guide, published by the McGraw-Hill Companies, and Means Residential Cost Data, published by the RS Means Company.

How Much Money Can You Make?

How much money can you make as a part-time remodeler? It all depends on the type and size of jobs you are doing and how well you manage your production schedule and financial budget. It would not be unreasonable to assume that you could make upwards of $40,000 a year as a part-time remodeler. Of course, the amount you make also depends on how much of the work you will do yourself and how much of it you will sub out. As with any projection, the amount of income potential depends on many factors, including the region in which you live. All in all, it is very possible to earn a full-time income on a part-time basis when you are a self-employed contractor.

CHAPTER TWO

When Builders Become Remodelers

Getting your feet wet in the building business without getting in over your head takes planning. Don't just decide one day to be a remodeler, run an ad in the local newspaper, and wait for the phone to ring. This is a big step in your career and not one to take lightly. A lot of planning and thinking needs to be done before that first nail is driven. To avoid failure, you must have some solid plans. Remodeling can be substantially different from building new construction.

PRO POINTER

The financial rewards in this business can be substantial, but remember that, where there is the opportunity for reward, there is also the potential to lose money.

The first thing you need to do is make sure that you are ready to assume the responsibilities that go with being a remodeler. Do you have enough general knowledge to perform the functions of a general contractor? If you don't, start reading, attending classes, or working on some construction sites. Gain as much experience and knowledge as you can before you offer your services as a remodeler to the public. If you have been a builder, your experience will go a long way in remodeling.

There are many ways to prepare yourself for becoming a general contractor. Read every book you can find on building, construction, remodeling, and related trades, and read books written for homeowners and do-it-yourselfers. Seek out titles that have been written for professionals, such as *Remodeler's Instant Answers,* published by McGraw-Hill. Absorb the wealth of knowledge provided by seasoned professionals that can be found in these books. Talk to people in the business.

You can also attend classes that pertain to various trades. Look into the possibilities of attending workshops or vocational classes if you feel you need more training than you can get from a book. Local community colleges frequently offer construction-related courses such as basic carpentry skills, blueprint reading, and even project management, and the classes are generally held in the evening and are relatively inexpensive.

PRO POINTER

You'll find many contractors who are willing to share some of their experience with you and point out some helpful tips to get you started as well as some pitfalls to avoid.

Videos have become extremely popular, and there are many available that show how to perform certain tasks, such as hanging cabinets or installing plumbing. Your local library or video store may have some of these learning tools on hand. If not, I'm sure they can help you order titles that will boost your skill level.

Go to some local residential work sites and check out what's going on. Make note of any subcontractors working there or names of companies delivering materials or equipment. This will help you become more familiar with the local construction community. Walk through some of the homes under construction and observe the types of materials and products being used. Watch some of the work being performed to gain a little more knowledge of that trade. I've found that you might be approached by the superintendent and asked what you are doing there, but once you tell them you are thinking about going into business for yourself, you'll find you will most likely be welcomed.

The business side of remodeling is critically important to your success. If you are not comfortable with your office skills, such as basic management and accounting principles, once again, investigate the business courses offered by your local community college. Now, assuming that you feel ready to become a remodeler, you must find a way to tap into this lucrative market.

PRO POINTER

Good administrative skills will make your entry into business much easier.

Which Type of Jobs Should You Start With?

Which type of jobs should you start with? Kitchens and bathrooms are usually the two best types of jobs to take on. However, if your experience is more in the field of new construction, garages and room additions can be a good starting point. Do jobs that you can do well to build a list of satisfied references.

If you are planning to work with experienced subcontractors, your choice of type and style of first jobs is broader, but logic dictates that simple jobs are faster and easier to complete. This is good for you when you are building references. It is reasonable that you should try to start with jobs that can be done quickly, so you can generate a cash flow and produce a profit as soon as possible. Keeping your starting jobs simple will make it easier to estimate material and labor needs, with less risk of cost overruns

Your Edge

What is your edge as the new remodeler on the block? It's something you have to create. It might be low prices or an outstanding design or superior workmanship at affordable prices. Something as obvious as marketing and advertising could be what sets you above your competition. The edge can be almost anything, but you need it to survive and prosper. If you are just a carbon copy of all the other remodelers, you will be at a disadvantage. Finding what will work best for you is a personal thing, but I can give you some ideas.

Price

PRO POINTER

Everyone enjoys receiving good value for their money, but some people view discounted products as damaged merchandise.

Price is a factor that many businesses use as a lever. Trying to beat the prices of your competitors would not be my first recommendation. If you become known as the cheap contractor or the discount remodeler, you will have trouble moving up to higher-priced jobs. But getting a reputation as a value-conscious contractor and a quality-oriented remodeler is a different story.

To create the aura of a value-based remodeler, you have to make your jobs a little different. Your goal is to make customers compare apples to oranges rather than apples to apples. This way your jobs don't appear to be a cheaper version of your competitors. Come up with a niche for yourself and build it.

Identify Your Customer Base

Before you make your selection of the types of jobs to do, you must identify with your customer base. Will you be dealing with old homes, more modern houses, or commercial properties? There is less money to be made on a per-deal basis with some jobs, but these entry-level jobs are a good place to begin your remodeling business. There are several reasons for this. Small jobs are often ignored by the larger, more successful contractors. This opens up the market up for you.

High-end jobs are lucrative, but you may find that you need a strong track record and good references to break into this work. Many new contractors concentrate on jobs that sell for under $10,000. This makes sense if you have limited financial reserves. It is risky to take large jobs that could cause you to go out of business if the job is delayed or you are not paid.

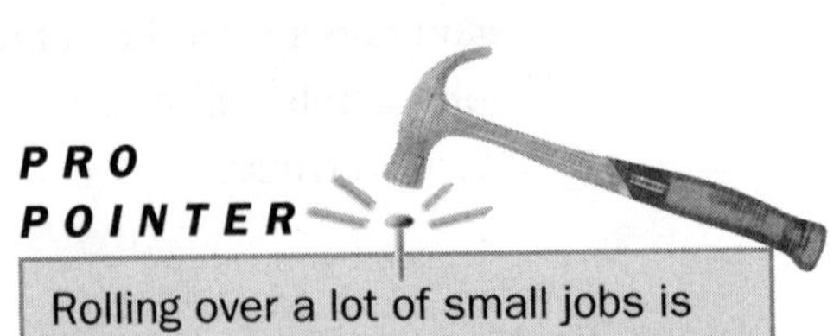

PRO POINTER

Rolling over a lot of small jobs is generally safer when you are getting started in the business.

Bringing It All Together

Bringing it all together to offer the public a fast, easy package is a sure way to success. Homeowners are usually excited but often naive. Most of them will respond to advertising, and almost anyone will listen to a sincere presentation from a caring, knowledgeable remodeler. You don't have to be the biggest contractor in town to capture your share of the market. But, you do have to be professional and persistent and have a reputation of honesty and integrity.

Let's say you have decided to run a few ads in the local newspaper and on cable television. Once you get some name recognition, you may decide to use a direct-mail strategy. There are companies that sell mailing lists depending upon geographic area, economic levels, family size, renters, homeowners, and so on. The mailing list you decide to use should consist of people who live in property that they own and who have adequate income levels to afford the work you plan to propose.

As a format and theme for your new company, let's say that you have decided to emphasize certain strong points in your offering focused on owners of older homes. To capture this market, you do research on the styles of homes that you hope to remodel. In doing so, you get a lot of design ideas. Use these ideas when crafting your advertisements.

When you start running your ads, you may be amazed at the response. Why are so many calling you, the new remodeler in town, when they could be calling established professionals? They're doing it because you identified a need, filled it, and made the public aware of what you were doing. I've done this type of thing time and time again.

PRO POINTER

If you know your market and can present your credentials well, you should find plenty of work.

Established remodelers may become complacent because they remain profitable and get enough work from word-of-mouth referrals so that they don't have to actively look for work. When homeowners approach some remodelers, they are not treated as potentially valued customers; I've heard this complaint from buyers on countless occasions. The homeowners feel as if the contractors don't want their business, and this group of customers is a prime target for your approach. When you are willing to talk to them as equals and offer advice on things to look for in remodeling their homes, customers will flock to you and spread the word to their friends about how great you are to deal with.

Test the Waters

When you are ready to get into remodeling, you have to test the waters to develop potential markets. If money is no object (and there aren't many of those contractors around), you can test the market by investing in newspaper or magazine advertising. For most people like me, research is the key. Talking with competitive contractors is a fast way to get some inside information. Visiting nearby residential developments may give you an idea of what is selling. Look to see what types of work is being done and who is doing it.

You will have to carve your own niche in the world of remodeling, and there are many of them, as we will discuss in later chapters. Mine has been kitchens and bathrooms. Some remodelers specialize in expansive, expensive room additions. The profit from larger jobs is good, but the volume may not be suitable for your needs. What works for me might not work for you. Everyone has to find their own place in life. Since I am a master plumber, kitchens and bathrooms are a natural for me. You might find that attic and basement conversions suit you better. In any event, remodeling is a good business to get into.

C H A P T E R T H R E E

Why Remodeling?

Whether you are thinking of opening your own business or trying to determine how to make your existing business better, you must know what it is that you want to achieve. One way to establish this is to list your goals and objectives. Business goals provide many advantages. They can pave the way to higher income and a more enjoyable life.

Going into business for yourself is a serious step to take and one that needs to be thought through carefully to avoid spending considerable sums of start-up money and energy, only to realize the business is not what you want. Faced with this kind of dilemma, there are two options open: shut down the business, which usually means losing all or a big portion of the start-up money, or continuing with a business that does not make you happy. Neither option is desirable.

PRO POINTER

Without goals, you have no direction. A business without direction is like a rudderless ship— bound to go nowhere.

PRO POINTER

Being self-employed requires discipline, long hours, dedication, and persistence. Owning a business is not the glorified cakewalk some people fantasize it to be.

It is not uncommon for people to put themselves into business situations that they regret. For some, the stress of owning and operating a business is too much. For others, the financial ups and downs are more than they care to deal with. Being in business for yourself is not all leisure time and big bank accounts.

While self-employment is rarely easy, it can be rewarding, both financially and emotionally. This chapter is going to show you how to use goals to reach the results you want–and you do need goals. Your goal may be to be successful, rich and happy, but you are not likely to make it without a realistic, step-by-step approach.

To say that you want a business with twenty employees is a goal, but you will have to arrive at that goal by a series of calculated, progressive steps. Unless you are very wealthy, you cannot afford to open your shop with twenty employees standing by to take care of business as it comes, because it may not come for some time. You must develop the business first, and the increase in employees will follow.

Have you ever noticed how some businesses flourish, while others falter? Maybe you have worked for companies that always seemed busy but never seemed satisfied with the profits or production. Why do you suppose this is? Have you wondered why some successful businesspeople are satisfied to keep their companies small, even when it appears that they could expand? The answers to these questions lie in goals or the lack of them. Let's take a look at how goals can affect you.

When Your Job Becomes Your Business

When your job becomes your business, your life changes. In fact, your job becomes your life–at least for a while. Everyone hopes the change will be for the better, but that is not always the case. You don't have a company supervisor to answer to, but you still have a boss. In fact, you now have many supervisors. Your new supervisors are your customers. If you don't do your job to the satisfaction of the customers, you won't have your new, self-employed job for long.

PRO POINTER

Owning and running your own business is not the same as going to your old job.

The truth is, being in business for yourself can be much more demanding than working for others. Let's say you are a carpenter. When you work for someone else, you have to worry about the quality of your carpentry work, showing up for work on time, and giving a fair day's production. After leaving the job and returning home, the rest of the day, evening, and night is yours. But when you own your own business, the work day does not end at 4:00 or 5:00 P.M.

When you are self-employed, you not only are expected to perform all the normal carpentry duties, but your work day will generally extend beyond the carpentry work. Paperwork must be done. Planning the next day's work schedule must be dealt with. Phone calls must be made and returned. Estimates may have to be prepared. Customer complaints and warranty issues must be taken care of. Marketing strategies may need to be developed and implemented. Accounts receivable and payable must be reviewed, calls made to collect money, and checks written to subcontractors and suppliers. The list for additional duties goes on and on.

PRO POINTER

Just focusing on some of these additional duties can highlight the difference between working on a regular job and having your own business.

Time with your family will be at a premium. Weekend outings may have to be canceled or delayed while you catch up on business matters that were not able to be completed during the week. Before you jump into the deep and sometimes turbid water of the self-employed, you should give careful consideration to your goals and desires.

PRO POINTER

Opening your own business is no small undertaking. The time and financial requirements of starting a business can be overpowering.

Construction Offers You Many Business Opportunities

The diverse nature of the construction business is what makes it so interesting. In the residential end of the business you can become a homebuilder or a remodeler. Some people, coming into the business on a part-time basis, become "Handymen," an ever-growing field where small jobs such as drywall, cabinet repairs, painting, and minor plumbing repairs are done quickly and yield high profits. The commercial side of the business is not only for shopping-center builders or office construction–a large part of this work is devoted to remodeling office space or small retail stores, again not very complicated projects. Many are short-term, where a good profit can be made. So when you are considering a career in construction, you've got a number of different scenarios to look at.

What Do You Want from Your Business?

What do you want from your business? This would seem to be a simple question, yet many people can't answer it. As a business consultant, I talk with a wide variety of people and businesses. When I go in to troubleshoot a business, the first question I ask is what does the owner want from the business? More often than not, the owners don't know what they want and give a broad, unfocused response to that question.

When I opened my first business, a plumbing business, I wanted to be my own boss. I wanted to work my own hours and not be worried about putting in 18 years, only to be let go before retirement. My dream called for building a powerful business that would take care of me in my old age. Well, I started the business and I was relatively successful. However, looking back, I can see countless mistakes that were made.

PRO POINTER

To be successful and to ensure the survival of your business, you must develop a business plan.

Since my first venture into business on my own, I have gone on to many new businesses, and each time I start a new venture I seem to find new problems to be dealt with. It is not that my methods don't work, but I always seem to find ways to improve upon them. I wouldn't begin to tell you that I know all the answers or can tell you exactly what you need to know to make your business work. But I can give you hundreds of examples of what not to do, and I can tell you what has worked in my business endeavors and those of my clients.

I don't believe you ever finish refining your business techniques. Even if the business climate is stable, you could always find ways to enhance your business. The business world changes frequently, forcing changes in business procedures and policies. If you are going to start and maintain a healthy business, you must be willing to change.

But back to the question, what do you want from your business? This question is applicable to people contemplating a business startup as well as to present business owners. Take some time to think about the question. Then write down your desires and goals on a sheet of paper. You need to write the goals and desires down. For years I refused to believe that writing my goals and desires on paper would make a difference, but it does. Someone once said that a thought is not a thought until it is spoken or written, and I firmly believe that.

With your list compiled, check over it. Break broad categories into more manageable sizes. For example, if you wrote down that you want to make a lot of money, define how much is a lot. Is it $30,000, $50,000, or $100,000 a year? If you jotted down a desire to work your own hours, create your potential work schedule. Will you work eight- or ten-hour days? Will you work weekends? Are your scheduled hours going to fit the needs of customers? This type of detailed planning improves your chances of success and happiness.

Where Do You Want Your Business to Be in Five Years?

Where do you want your business to be in five years? Most new businesses won't survive beyond their first year. By the third year, a high percentage of new businesses are defunct. A key step in securing a good future for your business is the development of goals and plans.

How big do you want your business to become? Do you want a big business for the sake of having a big business, or is your goal high income, which could be generated by a very profitable small business? How many employees to you want? Are you willing to diversify your business? These are typical of the types of questions you should be asking yourself. Let's take a moment to look closer at some thoughts for your business future.

Do you want a fleet of trucks and an army of employees? If you answer yes to this question, you must ask yourself more questions. Are you willing to pay the high overhead expenses that go hand in hand with a large group of employees? Do you have the management skills to deal with the increased complexity of a large business? Will you need to take classes on human resources to manage your employees? Will you have the knowledge to oversee accounting procedures, safety requirements, and insurance needs?

Almost every answer raises new questions. As a business owner, you must be prepared to answer all the questions. It is difficult for an individual to have the experience and knowledge to answer so many diversified questions correctly. For example, let's say you are about to hire your first employee. Do you know what questions you can ask without violating the employee's rights? Are you aware of the laws pertaining to discrimination and labor relations? What is your responsibility to that employee regarding insurance to protect against liability claims on the job, Social Security, worker's compensation, or unemployment? You could wing it and hope for the best, but that type of action could result in a lawsuit and a serious drain of your assets. You will need to consult with a professional in the field of human resources to answer these and other related questions. On second thought, maybe you should keep the business small and work it by yourself.

Do It Yourself, Sub It Out, or Diversify?

Are you willing to diversify? Should you consider hiring your own in-house trades and use them on the houses you remodel to save money? Could you subcontract the trades out to other remodelers when they need additional workers on their projects? It's possible that hiring your own plumber or electrician could make sense, but it is a big commitment and good tradesmen are hard to find. Often they come at a high price.

You could hire a master electrician and expand your business base. It would be appealing to have your own in-house electrician and the money made from service and repair calls outside your main business. So why not do it? If you don't know much about electrical work, you could be setting yourself up for long-range trouble by jumping into the unknown waters of electrical contracting.

PRO POINTER

There are many good reasons not to put a specialized tradesperson on your payroll. Overhead costs are certainly one good reason to keep your payroll roster lean.

How Big Is Big Enough?

How big is big enough? When you are planning the future of your business, you must use a measuring stick in your planning. When you think of the size of a business, what scale of measurement do you use? Do you think in terms of gross sales, number of employees, net profits, tangible assets, or some other means of comparison? Gross income is one of the most common measurement factors in a business. However, gross income can be very deceiving.

Theoretically, a higher gross income should translate into a higher net income, but it doesn't always work that way. Having a fleet of new trucks may impress people and create a successful public profile, but it adds tremendously to your overhead expenses and must be supported by a continued high sales volume. Having ten carpenters in the field may allow you to deposit large checks in the bank, but after your expenses how much is left? It has been my experience, and the experience of my clients, that you must base your growth plans on net income, not gross income. Determine how much money you want to make, and then create a business plan that allows you to reach your goal. Remember, you can always adjust your goal upwards as you become more secure in your business!

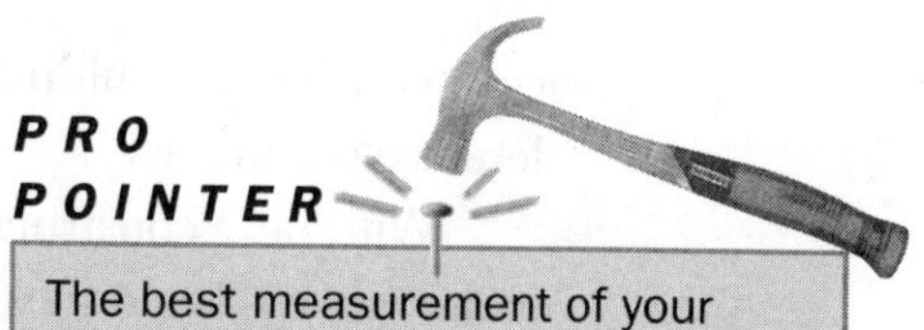

> The best measurement of your business is net profit, profit after all expenses have been deducted.

What Type of Customer Do You Want to Serve?

What type of customer do you want to serve? In the early stage of your business, any paying customer will be welcome. However, it is important for you to determine the type of clientele you wish to work with. The steps you make in the early months of your business influence the character of the business for a long time to come.

During the initial startup of a business it is easy to justify taking any job that comes along. The same is true when surviving in a poor economic climate. While this type of approach may be necessary, it should be temporary. For example, if you have decided to specialize in room additions, make every effort to concentrate on this field of work. When conditions do force you to do some remodeling or repair work, do it to pay the bills but continue to pursue your chosen goal. Let me give you an example from my past.

PRO POINTER

> If you bounce back and forth between different types of work, it will be more difficult to build a strong customer base and to streamline your business.

When I opened my first business, it was the only source of income I had. I wanted to be known as a remodeling plumber. After research, I had determined I could make more money doing high-end remodeling. I reasoned that remodeling was more stable than new construction and required less running around and lost time than service work. So I had a plan. I would become known as the best remodeling plumber in town.

During the development of my business, finding enough remodeling work to make ends meet was tough, so I took on some new-construction plumbing, cleaned drains, and repaired existing plumbing. It was tempting to get greedy and try to do it all, but I knew that wouldn't work, at least not with me being the only plumber in the company. Why wouldn't it work? It wouldn't work because of the nature of the different types of jobs.

With new construction, bid prices were very competitive. To win the job and make money, I had to work fast and efficiently. If I was plumbing a new house and my beeper went off, I would have to pick up my tools and leave the job to call the answering service and then call the customer. I then had two choices: to respond to the call or try to put the customer off until I left the new-construction work. Any way I looked at it, I was losing money on the house I was plumbing.

The time I spent picking up my tools and responding to service calls was eating into the narrow profit margin on the house. Service customers were annoyed if I didn't respond within an hour, so I began losing money and running the risk of making customers angry. If I left a remodeling job to answer a service call, the remodeling customer would become distressed because I left the job to take a service call. It didn't take long for me to see the potential for problems developing.

I began to set my sights on remodeling and put all my effort into getting remodeling jobs. In a matter of months, I was busy, and my customers were happy. My net income rose, because I had eliminated wasted time. In time, I added more plumbers and built a solid service and repair division. Then I added more plumbers and took on more new-construction work. But you must be careful in structuring your business plan, or it will get away from you and cause you to work harder and make less money.

When you consider your desired customer base and type of work, do it judiciously. If you live in a small town, you may not be able to specialize in only custom homes. You may have to do framing, roofing, or remodeling of all types of structures. You may have to do a little of everything to stay busy.

PRO POINTER

If you want to specialize in a certain field, never lose your perspective in pursuing the desired type of work and customers.

What Role Will You Play in the Business?

What role will you play in the business? For most entrepreneurs starting out, it is necessary to play all roles—worker, manager, and sales. However, just like setting a goal for the type of work your business will do, you should establish a goal for the role you wish to play in the established business. Do you want to work in the field and do your office work at night, hire someone to do it during the day, or work in the office and hire tradesmen for field work? Will you trust important elements of your business to employees, or will you want to do it all yourself?

The delegation of duties and proper management may be difficult for many first-time business owners because they may be totally unfamiliar with the complexities of running a business. Most people fall into one of two traps. The first group believes that they must do everything themselves, hesitating to delegate duties and, even after assigning the task to competent employees, constantly hovering over them checking their work. The second group believes everything can be delegated and fails to periodically check on employees' work. Somewhere between these two extremes is where most successful business owners reside.

It defeats the purpose if you hire employees to free up your time and don't allow them to do their jobs. You should always supervise and inspect the employees' work, but not by looking over their shoulder every five minutes. When your time is spent hovering over employees, you are neglecting many of your management and ownership duties.

PRO POINTER

If you did a good job in screening and hiring your employees, they should be capable of working with limited supervision.

If you decide to do everything yourself, you must recognize the fact that a time will come when you have to turn away business. One person can only do so much. It is better to politely refuse work than it is to take on too much and not get it done. That customer will appreciate your honesty and call you again or refer a friend to you. But when you take on more work than you can handle and consistently miss your completion dates, you have angered your customer and will not get a good reference on that job.

Will you be content to stay in an office? If your nature tends to keep you outside doing physical work, being office-based can be a struggle. It has its advantages, but office work can be a real drag on the person accustomed to being out and about and working with tools.

If you want to avoid office work, you can make arrangements that allow you freedom of movement. Answering services and an answering machine afford some relief. A receptionist is another way to keep the office staffed while you are out, but this is an expensive option. If you will be in the field, a pager and cell phone may be your best choice for keeping in touch with your customers. Devise the most efficient way to stay on the job and out of the office.

But suppose you want to be in the office. Who will be in the field? The solution to this problem is not easy for the new business owner to solve. When you hire employees for fieldwork, you will need to generate enough work to keep them busy. Getting steady work is tough, and for a new business, it can be nearly impossible. I have always run into the same problem: either I have too many employees and not enough work or too much work and not enough workers.

PRO POINTER

Don't overlook the issue of staying in touch and being readily available to your customers.

Most new businesses are run by their owners. If you decide that your place is in the field during the day, you must resign yourself to office work at night. But it doesn't have to stay this way forever. Decide where you want to be, in the field or in the office, and work up a plan to meet your goal.

Have You Evaluated Your Cash Reserves?

Have you evaluated your cash reserves? Any business needs cash reserves. A large number of businesses fail each year due to limited cash reserves. Without backup money, what will you do when a scheduled payment is due from your customer and is delayed? If you pay your bills late, your credit will be damaged.

How much money do you need in reserve? The answer to this question relates to the nature of your business and your ability to project and maintain budgets and schedules. The degree to which you wish to launch a marketing campaign is another factor. Advertising takes money, and new businesses need to advertise.

Some people say you shouldn't start your own business until you have at least one year's salary in savings. I must admit that this would be a comfortable way to get started, but for most people, saving up a year's salary isn't feasible. When I opened my first business many years ago, I borrowed $500 for tools and advertising. I had less than $200 in my savings account. Looking back, I was probably stupid to try such a venture, but I tried it and it worked. I'm not, however,

PRO POINTER

I have learned never to gamble more money than I can afford to lose, and I think this line of thinking is good. I wasn't always so cautious, but having a wife and children can change your perspective on worthwhile gambles.

Financial Statement

Your Company Name
Your Company Address
Your Company Phone Number

Date of statement: ____________________

Statement prepared by: ______________________________

Assets		
Cash on hand	$ 8,543.89	
Securities	$ 0.00	
Equipment		
2004 Ford F-250 pick-up truck	$14,523.00	
Pipe rack for truck	$ 250.00	
40' Extension ladders (2)	$ 375.00	
Hand tools	$ 800.00	
Real estate	$ 0.00	
Accounts receivable	$ 5,349.36	
Total assets		$29,841.25
Liabilities		
Equipment		
2004 Ford F-250 truck, note payoff	$11,687.92	
Accounts payable	$ 1,249.56	
Total liabilities		$12,937.48
Net worth		$16,903.77

FIGURE 3.1 Financial statement.

suggesting that you follow in those footsteps. I am a risk-taker. For me, trying a new venture is an adventure.

In evaluating how much money is enough, I suggest you develop a few worst-case scenarios. Draw up several different directions your new business venture might take—one that shows much less profit than you originally planned, one where your expenses exceed income but not by much, and one where expenses far outweigh income. You are looking for what could be the worst possible outcome of your decision to start or stay in a business—complete loss of your capital reserve. Once you have produced a number of these "what-if" situations, you will start to develop some insight into what you stand to lose. The next step is to determine how much you are willing or can afford to lose.

I don't think there is any clear-cut answer to how much reserve capital is enough. Each individual will have different needs. In my opinion, I would suggest having enough money to last at least four months without income. I would also suggest that when you feel you have established your monthly money need, add twenty percent to it because there will always be unforeseen expenses. If the business is not going well or your budget spending is running high, look for alternative sources of income.

PRO POINTER

Even with a four-month cash reserve, you must monitor the progress of your business on a regular basis.

The value of setting goals is known by most successful businesspeople. While it may seem silly to set goals, they make it easier to achieve desired results. It is important to separate your desires and goals. Setting standards that are too hard to obtain will only serve to disappoint and disillusion you. Attack your goals one at a time, and you will probably reach most of them.

C H A P T E R F O U R

Kitchens and Bathrooms

Kitchen and bathroom remodeling are great ways to get into the remodeling business. If you follow annual trends in magazines, you will see that bathrooms and kitchens are the two most profitable rooms in a home for homeowners to remodel. This should give you a hint. They are the two most desirable rooms to work in. I don't say this lightly. My experience spans about 30 years at this point, and kitchens and bathrooms have been the most lucrative.

There is more money to be made in kitchens, but the volume is in bathrooms. However, depending on where you are working, there are plenty of kitchens to go around, and they are very profitable. My experience has shown, and continues to show, that bathrooms are the number-one remodeling job in existing homes and that kitchens are also high on the list. If you have the right people, kitchen and bathroom remodeling can be very profitable. Let me give you an example of how I have my company set up.

I am not the average contractor. Remember, I've been doing this since 1974 and have been in business for myself since 1979. I started out as a plumber and was a master plumber and master gasfitter early in life. From there, I went to remodeling. Then I went to building 60 homes a year. This led me to become a real estate agent and property manager. I went on to become a Designated Broker and brokerage owner, but this is not the normal contractor's role. What follows is what you should strive to learn from.

By having the credentials that I do, kitchens and bathrooms are a natural match for me. Being a plumber, I fit right in. Additionally, I have a lot of related experience. I know how to do electrical work, but I am not licensed to do it for customers.

In reality, there is no phase of residential construction that I can't expose bad contractors in. This is another advantage that I have. The reality is that whether you have any trade experience or are just a good manager, there is plenty of profit in kitchen remodeling.

> **PRO POINTER**
>
> If you are a pure general contractor, you will have to pay subcontractors for a lot of the work that you could do if you had trade experience. This changes your profit potential, but it is still very good.

Magazines report every year how the two most valuable rooms in a home to invest in are bathrooms and kitchens. Based on my workload, bathrooms in my area far exceed kitchens. Admittedly, when I lived in Virginia, kitchens were more important than they appear to be in Maine. Either room is a good bet. Depending on which reports you read, kitchens often show a higher rate of return for homeowners than bathrooms do. But, both rooms are in the top two of what to remodel. As a general contractor, there is more money to be made in kitchens. This is due largely to the cost of the cabinets installed in a kitchen. Since contractors usually add a percentage of profit to the total job cost, there is more money. You can hardly go wrong with either room.

> **PRO POINTER**
>
> Your local market will drive your focus, but if you learn to specialize in kitchens and bathrooms, you should win.

There are multiple elements to kitchen and bathroom remodeling that you can capitalize on. However, you need the right people. Who are these people? The short list includes the following:

- Demolition crew
- Plumber
- Heating mechanic
- Electrician
- Carpenter
- Cabinet and countertop installer
- Drywall contractor
- Painter

- Flooring installer
- Tile installer
- Wallpaper installer
- Trash hauler

There can be other people needed, but those listed above are the primary people who are frequently involved in kitchen remodeling.

Will You Hire the Pros?

Will you hire the pros needed to do kitchen and bathroom remodeling, or will you sub it out? Your best bet is to sub it out, but it will not be an easy row to hoe. Finding good subs in the trades can drive you crazy. I'm lucky in that I know the game and can do it myself if push comes to shove. As a "Paper" general contractor, one who runs a business without being a hands-on worker, finding solid subs is an uphill battle that is lost more often than it is won. Expect to spend months to find the right people. You can do it, but the task is not simple or easy.

Hiring your own people is a major commitment. You must keep enough work coming in to keep them busy. The contracting business is often feast of famine. Throughout my years, the cycle has been similar from year to year. It's a tough business. What will you do?

PRO POINTER

It is typical to have too much work and too few people or too many people and too little work.

If you go with subcontractors, you don't have the amount of overhead expenses that you would have with employees. On the flip side, you will not make as much money on a per-job basis. It is a difficult decision. I have done it both ways and have had both good and bad results in each direction I took.

It has been my policy to keep "in-house" people to do what I call batting clean up. I put subs in to do the production work and then send in "my people" to fix their mistakes. If you plan to use subcontractors, you should either be able to do the job yourself of have loyal people who can do it for you who are on your payroll. If your only hope is subcontractors, you probably won't make it three years.

I like having my own people on payroll. The process gives you much more control. If you have the right people who show up everyday and who can do the work, you are way ahead of most of your competitors. On the flip side, you have a lot more overhead and paperwork to manage. This is an element that only you can make a decision on. Many contractors have been successful in both roles.

Getting the Work

PRO POINTER

Be very careful if you depend entirely on subcontractors.

Getting the work is a critical part of your business. This can be done in a multitude of ways. The best advertising that you can hope for is word-of-mouth referrals. But to get referrals, you need to get work to build a base of happy customers. The following is a list of some of my most successful marketing approaches:

- Direct mail
- Door hangers
- Newspapers
- Local periodicals
- Television commercials
- Radio
- Door-to-door solicitation
- Telemarketing
- Flyers
- Notices hung in local businesses and public bulletin boards
- Signs
- Imprinted clothing
- Seminars
- Newsletters
- Miscellaneous creative methods

Why Kitchens and Bathrooms?

I have already talked about some of the reasons for getting into kitchen and bathroom remodeling. People like nice kitchens and comfortable bathrooms and will spend substantial money to improve outdated fixtures and decor. They will do this for their personal satisfaction and for the potential resale value of their homes. This makes it easier for contractors to enjoy steady work and strong profits.

Some rooms in a home don't need to change much over time to keep up with current trends. And when they do, the improvements can be fairly minor. Use a dining room as an example. This type of room doesn't need a lot of attention to stay in modern condition. A light fixture might need to be changed. Walls will need to be painted and flooring may need to be upgraded. But overall, the room doesn't require a lot from a remodeling contractor. This is not the case with kitchens and bathrooms.

PRO POINTER

Since much of the money spent on kitchen and bathroom remodeling is usually recovered when a home is sold, homeowners tend to be comfortable with their investment.

Plumbing fixtures are redesigned frequently. What was popular for fixture choice ten years ago may not be desirable in current times. If you are old enough, you must remember some of the older colors of kitchen appliances. Do you remember Harvest Gold and Avocado Green? This type of refrigerator or range, while once popular, would be an eyesore today. This is one good reason to revamp a kitchen, but there are many more.

Kitchens and bathrooms can contain moisture. This is especially true of kitchens that do not have vented ranges hoods of an adequate size and power. The moisture can ruin walls, and this opens another opportunity for kitchen remodelers. Old tile on counters and floors can become faded, damaged, or just plain ugly. This also applies to old tile surrounding bathtubs or covering floors and walls. This is another element for the remodeling contractor to capitalize on. Light fixtures, cabinets, countertops, windows, and other design elements also offer good opportunities for remodelers. All in all, kitchens and bathrooms are an ideal place to gear up if you are a savvy remodeler.

Watch Out

What should you watch out for when remodeling a kitchen or bathrooms? There are all the normal considerations that range from rotted floor structures to uninsulated ceiling space and walls. Here are a few more common problems to beware of:

- Rotted flooring under toilets
- Undersized piping

Your Company Name
Your Company Address
Your Company Phone and Fax Numbers

PROPOSAL

Date: ________________

Customer name: ____

Address: ____

Phone number: ____

Job location: ____

DESCRIPTION OF WORK

Your Company Name will supply, and/or coordinate, all labor and material for the above referenced job as follows:

____
____
____
____

PAYMENT SCHEDULE

Price: ________________________________ dollars ($____________)

Payments to be made as follows:

____
____
____

All payments shall be made in full, upon presentation of each completed invoice. If payment is not made according to the terms above, Your Company Name will have the following rights and remedies. Your Company Name may charge a monthly service charge of _____________ (_____%) percent, ___________ (_____%) percent per year, from the first day default is made. Your Company Name may lien the property where the work has been done. Your Company Name may use all legal methods in the collection of monies owed to it. Your Company Name may seek compensation, at the rate of $_____ per hour, for attempts made to collect unpaid monies.

(Page 1 of 2. Please initial _______.)

FIGURE 4.1 Example of a proposal form. *(continued on next page)*

PROPOSAL (continued)

Your Company Name may seek payment for legal fees and other costs of collection, to the full extent the law allows.

If the job is not ready for the service or materials requested, as scheduled, and the delay is not due to Your Company Name's actions, Your Company Name may charge the customer for lost time. This charge will be at a rate of $________ per hour, per man, including travel time.

If you have any questions or don't understand this proposal, seek professional advice. Upon acceptance, this proposal becomes a binding contract between both parties.

Respectfully submitted,

Your Name
Title

ACCEPTANCE

We the undersigned do hereby agree to, and accept, all the terms and conditions of this proposal. We fully understand the terms and conditions, and hereby consent to enter into this contract.

Your Company Name — Customer

By: ______________________ ______________________

Title: ______________________ Date: ______________________

Date: ______________________

Proposal expires in 30 days, if not accepted by all parties.

(Page 2 of 2)

FIGURE 4.1 *(continued)* Example of a proposal form.

- Clogged drains
- The lack of a Ground Fault Interceptor Circuit
- Mold and mildew
- Adequate accessible opening for bringing in new bathing units
- Access to faucets and drains for bathing units
- The lack of independent electrical circuits for appliances
- Pathways for venting exhaust fans
- Adequate space in existing electrical services for new additions, such as a circuit for an island range top or whirlpool tub
- Split circuits and incorrect wiring that do not meet modern code standards
- Inadequate lighting
- Wet, damaged drywall
- Walls and floors that are out of plumb and out of square
- Space constraints for the remodeling effort at hand
- Delivery time for cabinets and countertops

Fast Turnaround Time

If you have the right people, there can be fast turnaround time in kitchen and bathroom remodeling. This is good for your cash flow. My crews can start and finish an average kitchen remodel in about 7 working days. This covers everything from complete demolition of an existing kitchen to full remodeling and cleanup.

In summary, kitchens are one of the best places to focus your remodeling efforts on. Once you become known as a kitchen expert, you should see a steady stream of interest in your services. I have specialized in bathrooms and kitchens for years and have never regretted it.

PRO POINTER

If you are working with a limited amount of reserve capital, the speed with which you can rollover kitchens makes this type of remodeling very attractive.

Custom Cabinets

Custom cabinets can add a lot to your profit zone as a general contractor. This is because they are expensive and you should be adding a percentage of the cost to your bid for managing the job. However, custom cabinets can take months to get. This must be a serious consideration. Many production cabinets are more than acceptable, and they are usually available on short notice. If you plan to sell custom kitchens, expect to wait a long time from the moment you sign a contract until you deposit the final payment from the customer.

Let's say that you have a 90-day wait for custom cabinets, and this would normally be a fast turnaround time. The cabinets arrive and something is wrong with them. It could take you six months just to have your materials to work with. Whatever you do, don't demo out a kitchen until you have all the replacement parts, including cabinets, in hand and inspected. Leaving customers with a kitchen sink propped up on a makeshift cabinet is not a good way to get started in kitchen remodeling. Have I had this happen? Yes. In fact, it happened just recently. Let me tell you the horror story.

PRO POINTER

Production cabinets that are of high quality will get the job done quickly, and they may not cause the problems that may be equipped with custom cabinets.

Warped Cabinets

We recently had a major franchise supply company come to our area and sold one of my customers warped cabinets. This store has provided many of our customers with substandard materials. The public is not happy. This company has eaten into our profits for the moment by promising a lot for a little, but customers will wise up soon. Anyway, here's the deal.

My company went to replace a kitchen about 90 days ago. The owner-supplied cabinets were boxed in the homeowner's garage. We were told to rip out the kitchen, and we did. When my cabinet installers brought in the cabinets and started to install them, it became clear that the cabinets were junk. Not only did the doors look like they were warped enough to resemble bananas, the boxes of the cabinet structure were warped and some of them were not even sized properly. The job was a disaster. I rigged up a makeshift sink arrangement for the customer to get them through until the new cabinets were delivered.

The second shipment of cabinets came in about a month later. They were not the right layout and they were the wrong color. So, the cabinets had to go back again. Then, some four weeks later, the next replacement shipping arrived. Was it better? Yes. Was it right? No.

After about 90 days, the customer had become so frustrated that they were willing to accept less-than-perfect materials, at least for some of the cabinets. The defects cost my crews time to correct them, but we did it to help the homeowner. Essentially, the kitchen became usable in one day, but there is still one cabinet that has to arrive for us to install and some incorrect trim that was shipped. The countertop was warped and there is still a section that the supplier has to replace for us to install. What a mess.

My people made the best of a bad situation, but the customer suffered from the supplier, and they are still suffering. There is no real excuse for this. While the cabinets were not of good quality, they still cost over $6,000. If the customer had hired me to take care of the cabinets, this would not have happened. One of the biggest problems with kitchen remodeling is cabinets and countertops, and you have to stay on top of it. But, custom cabinets may not be the way to go for an average kitchen. Consider a high-grade production cabinet as an alternative. If you are doing an extremely custom kitchen, then a cabinet shop is your best bet.

A Cabinet Facelift

A cabinet facelift is a cheap way to give a kitchen a new look. This is basically when you replace cabinet doors and exterior finishes without removing the existing cabinets. I have limited experience in this field, since I have never taken this path. Due to my lack of experience in refacing, I cannot comment with authority on it. In theory, the concept works, but that is the most I can tell you on this subject.

Flooring

What type of flooring should go in a kitchen or bathroom? The answer depends on the style of the room. Sheet vinyl is the most common type of flooring used in a kitchen. Tile is another suitable selection, but it should be a type of tile that will not be slippery when grease or water spills onto it. Some customers prefer wood flooring, and this is okay for a kitchen. I would stay away from wood flooring in a bathroom, but it is okay for a kitchen. Carpeting is acceptable, but probably not a good idea.

Countertops

Countertops are a major element of kitchen remodeling. Most counters are preformed tops. These are fine. Some countertops are made out of granite. These are heavy and expensive. Tiled countertops are not uncommon, and they are very nice.

Then there are other types of specialty tops that give a special look to a kitchen. All of these choices are fine. You have to match the countertop to your customer's needs, desires, and budget.

Garbage Disposers

If your customer wants a garbage disposer installed, check with your plumber for local code requirements if you work in an area where private sewage disposal systems are in use. Some code jurisdictions don't allow garbage disposers to be installed in kitchens that discharge into septic tanks.

Light

Most people agree that they want a lot of light in their kitchen. My company installs skylights, garden windows, and similar improvements to increase natural light in kitchens. Additionally, you can add under-cabinet lights, recessed lights, and track lighting to generate artificial light. Most people will agree that a bright kitchen is a happy kitchen.

Homework

Do your homework before you begin selling kitchen and bathroom jobs. Read books on design and remodeling elements. Then read some more of them. Get fully immersed in the topic. Once you do, the profit can be substantial.

Customers need guidance. You might consider finding some kitchen planners to work with if you don't feel comfortable giving customers options. There is potentially a huge amount of money and customer satisfaction to be made with kitchen remodeling. Don't overlook this opportunity. You can do it, if you are willing to educate yourself and devote your attention to this type of remodeling.

PRO POINTER

There are so many options for kitchens and bathrooms that you will have to work a bit to become an expert in the field.

CHAPTER FIVE

Setting Up Your Business

Taking the plunge into your own business is a big financial responsibility. There's really no way around this fact. You simply can't become a professional homebuilder without running the risk of losing money. While some of the risks are basically unavoidable, there are many business expenses you can curtail, and by trimming the "fat" where you can, it's possible to lessen the odds of failure for your business.

Remodelers can get caught up in the large amounts of money that they handle. It's not uncommon for a contractor to make bank deposits in excess of $50,000, but remember that most of that money is not yours. This money deposited into a contractor's operating account is there to pay suppliers and subcontractors, not for the remodeler to spend on personal items.

To reap the highest profits, it is essential to keep a cap on all expenses. On the other hand, it is possible to be so consumed with keeping expenses low that you damage your chances for big success. So you have to know which expenses to cut out and which to keep.

PRO POINTER

Some remodelers see money and become extravagant in their expenses—a sure-fire way to destroy your business.

Keeping Your Expenses in Check

Keeping a firm grip on expenses and keeping your business plan in mind will help you reach your goal–being a successful contractor. Times change, and businesses must change with them. Your time is valuable and must be spent wisely. How you handle your time and your money will create the difference between success and failure.

Remember, for every dollar you save in expenses, the higher your profit will be. However, cutting the wrong expenses can cost you much more than what you save. How will you know which expenses to cut? Can you predict the future? Not really, but by looking back over economic events of the past you will become more familiar with certain patterns–boom and bust and lots of situations in between. While you may not be able to tell which horse will win the next race, you can make projections for your business that will be more fact than fiction.

Beating Heavy Overhead Expenses

Beating heavy overhead expenses is one of the best ways to ensure the success of your business. Excessive overhead can drive you out of business. If you fail to scrutinize your operating costs, you may find yourself out of business and looking for a job. Since this is such an important aspect of your business management, let's take a closer look at how you can get a handle on your overhead expenses.

PRO POINTER

Overhead expenses are expenses that are not related directly to a particular job. Examples include rent, utilities, phone bills, advertising, insurance, office help, and so on.

Rent

Office rent may not be much of a factor in your business. If you work out of your home, you will not notice any new financial strain by converting one of your rooms into a designated office. In fact, you will be able to write off a portion of your household expenses, mortgage, taxes, and so forth. If you are considering a home office, talk to your accountant to find out what these deductions mean to you. If you rent a commercial office space, you will incur increased money demands. Don't forget that landlords generally increase the rent yearly.

How much you pay in rent will depend upon how much space you lease and where the space is located. Prime locations are expensive for an office of any size. While you might rent a small upstairs office in an average location for less than $600 a month, the same space in a fashionable part of town could cost upwards of $1,000 a month. The size of your office will also affect its cost. And, location has a major impact on office expenses.

Should you rent or just work out of your home? A home office doesn't get you away from your family and household. This can be a disadvantage, especially for people lacking self-discipline. It can be easy to leave your home office to take care of that household chore you put off last Saturday. It is equally easy to leave your desk to watch your favorite show on television. If your spouse or children come into the office, you are likely to lose production time. Background noise can be distracting when you are talking business either in the office or on the phone. But on the other hand, let's say you have a free hour at night or on the weekend. You can just go to your home office and put in some work reviewing weekly expenses, working up an estimate, or planning your week.

PRO POINTER

If you can operate your business from your home, you can save the cost of rent, but there are sacrifices.

If clients will be coming to your office, a home office can have a negative effect on the perception of your business success. However, many customers will envy you for being able to work from home. In general, a home office is a good idea for a new business, and it can be inexpensively furnished to project a degree of professionalism.

If a home office is not for you, you will have to turn to commercial space. With rent being an overhead expense, you need to determine how much you can afford to pay. What did you allocate in your business plan? Does that figure still hold true, or are you under budget for several other overhead expenses and able to spend a little more on your office rental?

PRO POINTER

Your relationship with the real-estate agent is another marketing tool because your contact may keep you in mind when another client asks about finding a builder.

If you have developed a well-thought-out business budget and done your business forecasting homework, you can make some assumptions as to how much rent is too much. If you must rent commercial space, you could decide how much to spend by seeing what's available on the market. You should know how much space you need and what locations are acceptable to you. With this knowledge, shopping for an office is a simple matter. You simply respond to various for-rent ads and check prices and amenities. You also ought to consider contacting a real-estate agent who specializes in small-office rental space.

Once you know what office space is renting for, you are in a position to make a decision. Some of the deciding factors might include the term of the lease, the amount of the security deposit, whether utilities are included, whether a security system is included, and other expenses related to the office.

For most contracting firms, a plush office is not necessary. As long as the office space is clean, well organized, and accessible, it should be fine. In deciding on rental overhead, keep your costs as low as you can while still getting what you need, no more and no less.

Utilities

Utilities are one overhead expense you can't do without. Heat, hot water, electricity, air conditioning, public water fees, and sewer fees need to be controlled. Using night setback thermostats, keeping air conditioning to a minimum, and replacing incandescent light bulbs with fluorescent bulbs are just some of the ways you can control utility expenses.

Phone Bills

Phone bills for a busy business can amount to hundreds of dollars each month. While it is often imperative to make business calls, you may be able to trim your phone bill into a more manageable range. Shop around for the best deal—but don't forget to read the small print. I recently changed my service for unlimited local and long distance calls based on a $49.95 per month ad—when the bill came through it was for $65.00. (The ad neglected to mention all of the various and weird fees, taxes, and so forth.)

Take advantage of all the discount programs offered by the many phone services. Another possibility is to make your long-distance calls after normal business hours. Reducing idle chitchat can have a favorable impact on your phone bill. If you have directory advertising billed on the invoice for your telephone service, you might consider altering the size of your advertisement.

PRO POINTER

Don't be too quick to cut back on your directory advertising. You may lose more business than the savings are worth.

Advertising

Advertising is a must-do expense for most businesses. However, you can make your advertising dollar stretch further by making it more effective. Keep records on the pulling power of your ads. Track your responses to the ads and the number of responses that turn into paying work. Target your advertising to bring in the type of work that is the most profitable for your firm. By refining your advertising, it will pay for itself.

Insurance

You cannot reasonably avoid the burden of insurance. There are two keys to controlling your insurance expenses. The first is to avoid overinsuring the company. There is no point in paying premiums for more insurance than you need. The second key is to shop rates and services. Insurance is a volatile market; the rates change often and quickly. The insurance you had last month may need to be reassessed this month. By doing periodic evaluations, you can maintain maximum control over your insurance expenses.

Try to find a reputable insurance agent that has had experience with contractors and discuss your plans. Again, your agent is another potential source for future business, as he or she might refer another client to you. And, when a bond is required, your insurance agent will be able to assist you.

Professional Fees for Lawyers and Accountants

Professional fees for advice or consultation with a lawyer or accountant will more than likely not occur monthly, but these fees can amount to hundreds or thousands of dollars a year. If you have a CPA do your taxes, and you probably should, you can save money by doing some preparation work. By having all your documents organized and properly segregated (each overhead expense in a separate folder and labeled), you will save the CPA's time. By submitting a complete tax package to the accountant, you will reduce the number of phone calls, visits, and time you spend with the professional. All the time you save for the accountant will result in a lower fee.

PRO POINTER

Certainly, you should engage professional help when you need it, but you can lower the cost of these services by doing some of the work yourself.

When looking for an attorney, look for one who has contractors as clients. He or she will be more familiar with your work and will have a ready file of forms and contracts that have been developed and proven successful over the years. Attorneys should draft and review legal documents. However, if you know what you want, you can reduce time spent with them. Since they generally bill on an hourly basis, you will save money if you prepare a draft of the document you want before meeting with them. When you meet with your attorney, have your questions prepared, preferably in writing and organized. The quicker you get in and out, the less you will have to pay.

Office Help

Your business may or may not require office help. If you have personnel in your office, they are an overhead expense. Employees can be one of your most expensive overhead items. Avoid this type of overhead expense until your business can't function properly without it.

Office Supplies

Office supplies may not seem like a large expense, but they can add up. One way to reduce the cost is to buy them in bulk. Instead of purchasing one legal pad, buy a case. Buying in bulk and buying from wholesale distributors can reduce the money you spend on disposable office supplies.

Office Equipment and Furniture

Every office needs some office equipment and furniture. You will need a desk, a chair, and a telephone. You should have a filing cabinet, and you will want other pieces of furniture and equipment.

Be selective in what you purchase. Before you buy anything, make sure you need it and that the cost is justified. Office equipment is an area where many business owners go overboard. If there are used-office-furniture companies in your area, pay a visit to one of them and see what bargains they may have.

Do you need a copier or fax machine? With prices of electronics dropping drastically, you can now purchase a combination copier/fax machine for under $100. Even high-quality laser copy machines can be purchased for $250 or less. Shop around for deals in office machines and consider the cost of replacement cartridges and other replaceable accessories when evaluating the initial and operating costs of the equipment.

Do you really need a computer? It comes in handy when writing letters or proposals, and with the wide range of software on the market, this versatile machine can keep your books, help design a room addition, become a small-scale printing service, and produce attractive flyers and handouts. And with an Internet connection, you can tap into on-line estimating services, locate special products, and seek out technical and industry information.

But don't get caught up in the bells and whistles—an adequate desktop computer can be purchased for under $700, but with the addition of flat panel monitors, wireless connections, and other add-ons the cost can go beyond $2500. After you have been in business for a while and start to feel more secure in your ability to control your operations, you can look into purchasing a more versatile computer.

Vehicles

A company vehicle is an overhead expense. While it's true that you need transportation, you don't have to buy a loaded, top-of-the-line vehicle. If you can do your job in a $9,500 mini-pickup truck, don't buy a full-size truck at twice the price. Cutting your overhead is a matter of common sense and logic. Buy what you need and don't buy what you don't need.

Learn What Expenses to Cut

You must learn what expenses to cut. Cutting the wrong expenses may be worse than not cutting any expenses. We have just finished looking at normal overhead expenses. What other expenses can be cut back?

Field Supervisors

Hiring a field supervisor will add significantly to your overhead. Not only will the supervisor's salary be an added cost, but there are benefits that must be added to the basic hourly rate. Known as the "labor burden," employer's contribution costs such as Social Security, currently at 7.5% of wages, plus worker's compensation and state unemployment tax can really add up. If you currently have a field supervisor, ask yourself if you could do the field supervision and use that individual in another capacity, possibly as a working foreman. There is a balance between a business owner who is unwilling to delegate duties to anyone and an owner who puts too much responsibility on employees.

You will gain many advantages by being your own field supervisor. You will see firsthand how your jobs are progressing. Customers will see you on the job and be more comfortable that they are getting good work and special attention. The cost of having an employee as a supervisor will be eliminated or reduced. Before you plan to hire a field supervisor, consider doing the work yourself.

PRO POINTER

Field supervision is mandatory for many contractors, but unless there are numerous jobs going simultaneously, you as the owner should be able to supervise the fieldwork.

Surplus Materials

Surplus materials are common in the contracting business. Since the quantity of these leftovers is usually minimal, many contractors store them in their office yard if they have space or in a rented area. If the items will be used within a month, putting them in storage is not a bad idea. However, if you don't know when you will have an occasion to use the materials, return them to the supplier for credit.

By returning surplus materials, you free some cash that was previously tied up in unneeded inventory, and you have no need for a large storage area. When the materials are moved from the job to your storage facility, this double handling becomes expensive, and in the end the cost just to move these materials may exceed their purchase price.

Most suppliers that you deal with regularly will be happy to pick up your leftovers and credit your account. By taking this route, you are not paying to have the materials removed from the job site, and when you need the materials again, the supplier will deliver them for you. You save money both ways.

Travel Expenses

You can cut your travel expenses by keeping mileage logs for all your vehicles. When tax time comes, you can deduct the cost of your mileage from your taxes. The amount you may deduct is set by the government, but it is enough to make keeping a mileage log worthwhile. Talk to your accountant. Since there are two ways to track vehicle expense—by mileage (the government allows so much per mile) or by actual expenses (insurance, gas, oil, and repairs). Find out which system is best for you, remembering that once you adopt one system for a vehicle, you can change later if circumstances change.

Cash Purchases

Most contractors have occasion to purchase small items with cash. The items might be nails, photocopies, stamps, or any number of other business-related items. By keeping the receipts for these items, they can become tax deductions. While a receipt for less than a dollar may not seem worth the trouble of recording, if you collect enough of them, they can add up to a rather large amount.

Periodically Review Your Expenses

You can decide which costs to cut by periodically reviewing your expenses, such as truck repairs, office expenses, and utility costs. Do the expenses incurred to date seem reasonable; how do they compare to your budget? Was your budget unrealistically low or high?

Before you incur or cut an expense, be sure you are doing the right thing. Taking action too quickly can result in costly mistakes. For example, canceling your telephone-book advertisement will probably be a regrettable mistake. Dropping your health insurance may come back to haunt you. You must scrutinize all your expenses and cut only the ones that will not hurt you or your business.

Know Which Expenses Are Justified

You should know which expenses are justified and should not be cut. Just like your ad in the phone book or your health insurance, some expenses are absolutely necessary to keep your business going.

How will you know which expenses are justified? If you rethink the amount spent and the reason for incurring the expenses, you will be able to tell if they are necessary. For example, a plumber would be lost without a right-angle drill. Some tool expenses are more than justified; they are necessary because they provide savings in labor. A painter's ladders would fall under the same classification. These expenses are obviously needed.

The difficulty comes with costs that are not so obvious. Does a plumber need a backhoe? There are many times when plumbers work with backhoes, but the occasions usually are not frequent enough to justify purchasing the equipment. However, if the plumbing company does extensive work with a backhoe, it might be wise to purchase or lease one. You might consider yourself to be in a similar situation. As a builder, you no doubt have many occasions when a backhoe or excavator is needed.

Does this mean you should rush out and buy one? I don't think so. You will normally be much better off to hire a subcontractor to do this type of work for you. If you or one of your employees is a trained operator, rent one for the day and line up enough work to make a full day's rental pay off.

If you computerize your office, you will need a printer. Look at ink-jet printers and laser printers—both of which can be purchased for under $300. Inquire about replacement cartridge costs to determine which one will be less expensive to operate. Maintenance contracts may not be advisable if you buy a well-known brand with a history of reliability.

Most business owners will agree that advertising is a necessary expense. But don't spend money on ads that fail to produce any response. Quality is worth more than quantity. Some ads pull large responses but net few jobs. Other ads pull fewer responses but result in more work.

Don't place ads in every medium available. You are wasting money. Advertisements should be keyed to let you know which ads are working and which are not. The odds are good that you can trim some of your advertising expenses without losing business. Get some help in formulating your ads. Newspapers and magazines often offer assistance in putting together an effective ad. Look at other remodeler's ads and pick out some of the good points and attractive layouts.

PRO POINTER

If you don't reduce or control nonessential spending, you will reduce any potential profits. If you reduce or eliminate the wrong expenses, you could risk losing business.

You need to walk a thin line when cutting expenses. If you don't reduce or control nonessential spending, you will reduce any potential profits. If you reduce or eliminate the wrong expenses, you could risk losing business. Knowing which expenses to cut must be learned with testing and experience.

Cutting the Wrong Expenses Can Be Expensive

Cutting the wrong expenses can be expensive. If you cut the wrong expenses, you may lose much more money than you save. Some of your actions may be difficult to reverse. For these reasons, you must use sound judgment in making cuts in your business expenses. Let's look at some of the expenses that you may not want to cut.

Directory Listings

Directory advertising in the phone book is one of the first expenses many business owners contemplate cutting. Before making cuts in your directory advertising, remember that you will have to live with the change for a full year. If you reduce the size of your ad or eliminate it, you can't reverse your actions until the next issue is printed.

Directory advertising does work. It is a proven fact that people do let their fingers do the walking. Remodeling contractors can often do well with a much smaller ad than a service and repair company. People looking for remodelers are not dealing with an emergency situation, as they might be with an electrical or plumbing problem.

Listings in the local phone book lend credibility to your company. They prove that you have been in business for a while and show that you are a local concern. The size of your ad in the directory can be influenced by your competition. If other builders are running big ads, it may be necessary for you to follow suit. However, many people believe that a box ad in the column listings is sufficient. You should remain listed in the phone directory in one way or the other.

If you track your phone inquiries to determine which advertisements they are calling from, you can determine what percentage of your business is coming from the phone book. I am sure that you will get calls from directory advertising, but the cost of the ad could be more than the calls are worth. Track your calls for a year and decide if you are getting your money's worth from the phone book. It may be wise to have a modest listing in the directory and spend the money you save on a more targeted form of advertising.

PRO POINTER

The main thing is not to make radical changes in your directory advertising before you are sure they are justified. A year is a long time to live with a mistake that hurts your business.

Answering Services

Human answering services are another frequent target of expense cuts. While some businesses do all right with answering machines, most businesses do better when a live voice answers the phone.

Most answering services can be terminated and picked back up the following month. If you are unsure of the value of your answering service, terminate it for a month and compare the number of leads you get for new work. If you don't notice a drop in business, you made a wise decision. If, however, you are losing business, you can reinstate the answering service.

Health Insurance

Health insurance is very expensive, and many business owners consider eliminating their coverage at one time or another. You are taking a big risk to drop your health insurance. As expensive as the insurance is, if you have a major medical problem, the insurance will be a bargain. The accumulation of big medical bills could drive you into bankruptcy.

If you feel you have to alter your insurance payments, look for a policy with a higher deductible amount. Such policies have lower monthly premiums and still provide protection against catastrophic illness or injury.

Dental Insurance

For most people, dental insurance is not as important as health insurance, but it is still a comforting thing to have. If you have bad teeth or need crowns and root canals, dental insurance can pay for itself. There are, of course, other types of dental services that will make your insurance premiums seem small. Try to avoid cutting out any of your insurance coverage.

Inventory

While you shouldn't carry a huge inventory that you don't need, you should stock your trucks with adequate supplies. Check your inventory from time to time to determine if you are stocking the right items or have bought materials that haven't been used in six months or so. If you cut back too far on inventory, your crews will waste time running to the supply house, and your customers will become frustrated by your lack of preparation for the job.

Retirement Funding

Retirement plans are frequently one of the first expenses cut by contractors in money crunches. A short moratorium on retirement funding is okay, but don't neglect to reinstate your investment plan before it is too late.

Bid-Sheet Subscriptions

Bid-sheet subscriptions are sometimes put under the financial microscope. While these expenses are not monumental, they appear to be an easy cut to make. If all you do with your bid sheets is glance at them and trash them, by all means cancel your subscription. If, however, you bid work on the sheets and win some jobs, eliminating your subscription could be the same as turning work away.

Credit Bureaus

The fees charged by credit bureaus can become a target for company cuts. Before making the decision to do without credit reports on your potential customers, weigh the risks you are taking. Doing work for one customer who doesn't pay would more than offset the savings you made by eliminating the credit bureau fees.

Advertising

Advertising is a tricky topic. When business is off, you need to advertise to get more business. But when business is down, so is your bank balance. Justifying the cost of advertising when your money machine is running on fumes is not easy. It is not wise to eliminate your advertising, but it is smart to target it. Do a marketing study to determine where and what to advertise, and then advertise. If your marketing research is accurate, the cost of your advertising will be returned in new business.

Sales Force

When times are tough, business owners look to the sales force. Even when the salespeople are paid only by commission, some business owners consider eliminating them. This makes no sense to me. If you have a sales force that only gets paid for sales made, why would you want to get rid of them? Unless you are going to replace the existing sales force with new, more dynamic salespeople, the move to eliminate sales doesn't make much economic sense.

Looking Into the Future

You don't need a crystal ball to look into the future. What you need is determination, time, and skill. Time can be made and skills can be learned, but you must already possess determination. If you aren't motivated to look toward the future of your business, you will find it difficult to do so. On the other hand, if you are committed to making your business successful, you can do a fair job of projecting your future in business.

Your business will face many challenges over the coming years, and you must be prepared to deal with them. With adequate preparation and thought about where you are going, your business has a good chance of lasting a lifetime. In order to be prepared, you must start now.

How can you judge what your business will encounter three years from now? Our economy runs in cycles. The businesses of most contractors are affected in some way by the real-estate market, and the real estate market is greatly influenced by interest rates. When rates are low, as they were during the first part of 2000, a flurry of refinancing took place, and people found that they could not only afford to buy a house but could get more house for their money because of lower monthly payments. During interval periods, many companies make moves to improve business but often fail to survive the down periods of the economy.

Since the economy is cyclical, past history can give you a glimpse of what may be around the corner. The clues you find may not be right on the money, but they are likely to render a clear picture of what's in store for your business.

PRO POINTER

Construction and remodeling historically has been subject to "feast or famine" cycles.

In the early 1970s, the real-estate market was booming. People were making and spending money. By the late '70s and early '80s, the business environment took a nosedive. Interest rates soared and business production in most fields dropped. Contractors scurried to collect money due them and to find new work. Times were tough, to be sure, but many contractors made it; I was one of them.

By mid-1980s, business was good again. I was building as many as sixty homes a year, and most contractors were expanding their businesses, myself included. As time passed, the economy started shifting to a downward turn. By the late '80s, the business world was reeling again. Once again, the economy was sagging and businesses were closing their doors. From the '90s to the early 2000s homebuilding in many parts of the country was booming, brought about primarily due to mortgage rates in the 5% to 6% range. What can you learn from this abbreviated lesson? You might see a trend for major slow-downs in the economy at least once in every decade. You might also assume that financial failures over the last twenty years have occurred more often in the latter part of each ten-year period.

If you looked deeper into the history of the last twenty years, you would find some interesting facts. In the late '70s and early '80s, banks were quick to rise to the problems at hand. Creative financing blossomed, and the business world turned itself around. In the more recent recession of the late '80s and early '90s, the banks did not rally to help. Instead, many of the local banks were bought out by larger ones. Business owners may not have high interest rates to contend with during a recession, but they also may not have willing lenders to help them with their financial battles.

PRO POINTER

You may have to face a recession within the first ten years of your business.

The state of employment, or should we say unemployment, during the period 2000 to 2004 is another factor in the economic cycle. More and more high-paying jobs are disappearing, and families often have to have two wage earners in order to survive. If interest rates start to climb sharply in the next few years, the robust building cycle might come to a screeching halt. All these factors need to be part of your thinking as you make plans to push ahead with your business.

There are different types of economic wars to win. If you have recently gone into business, you know what you may be dealing with—consumers who are taking advantage of low interest rates to buy a new house or remodel their existing home—but what will be the outlook if the current rates climb and more unemployment occurs? When times are good, that's the time to draw up contingency plans in case the economy takes a dip and work is scarce.

Reading old newspapers at the library is one way to revisit the past. Talking with people who have lived through the tough times is another good way to gain insight into what happened and why. Tracking past political performances can produce clues to the future. The key is spending the time and the effort to look back and understand why good or bad times occurred so that you can look ahead and be somewhat prepared for the future.

Once you are familiar with the past, you can see trends as they are forming. You will be able to spot danger signals. These early-warning signs can be enough to save your business from financial ruin.

PRO POINTER

Unless you have an astute business advisor, you will have to learn to pick up early-warning signals on your own.

Your market study can be compared to storm trackers; certainly more people have been spared the pain of these vicious storms since scientists have studied and tracked past storms.

Long-Range Planning Pays Off In the End

Long-range planning pays off in the end. By preparing for the worst, you can handle most situations that come your way. Your plans must focus on financial matters, but that's not all you have to plan for. As you grow older, how will your business be run? If you do your own fieldwork, how will it get done when you are no longer physically able to do it? How quickly will you allow your business to grow? These are only some of the questions you will want answered when planning for the future.

If you ignore the changes that are taking place around you, you will be unable to cope with what the future will bring. Can you imagine how business owners of only a few years ago felt about computers? Do you think they ever dreamed that computers would have such dominance in the marketplace? Change is inevitable and beyond your control, but planning for change is within your control. To have a long-term business, you must have long-term plans.

Focusing on Change

Positioning your company to meet changing economic conditions and demands will require your attention. Businesses don't change themselves. People change them, people just like you. You may not be focusing on this direction now, but you should start making plans for the changes soon.

Growing Tired of Working in The Field?

As the years go by, you may get tired of working in the field. What used to be fun may not be so enjoyable at an older age. The physical work that has kept you in good shape may become a bit much for you in ten or twenty years. How will you adjust for the desire to get away from the physical work?

Some contractors plan to stay in the field until the day they retire. Many anticipate hiring employees or subcontractors to pick up the slack in the physical work. Of the two options, I would recommend planning on hiring help. You may well get tired of working in the field before you can afford to retire.

If you intend to bring employees or subcontractors into your business at some point, you can start planning for the change now. In your spare time read books about human resources and management skills. If there are any courses offered in your area, take the time to attend and learn the skills you need to operate as an effective manager. The knowledge you gain now will be valuable when you enlist the help of others in your business.

Company Growth

PRO POINTER

Without a plan for business growth, too rapid growth with no preparation can be as deadly to your business as losing sales volume.

Company growth is a concept that many business owners ignore because their energies are focused on the day-to-day task of running their business. Many owners go about their business, and when volume increases don't know how to deal with it. This helter-skelter method is no way to expand a business. I know it may seem strange that having a bigger business could be worse than maintaining your present size, but it can.

When owners allow their companies to balloon with numerous employees, subcontractors, and jobs, management can become a serious problem. This is especially true for business owners with little management experience. The sudden wealth of quick cash flow and more jobs than you can keep up with is a company killer. Increased cash flow coming in will probably mean increased cash flow going out, and if one big customer is late in paying, will you be able to continue to pay your bills on time?

The operating capital that kept your small business floating over rough waters will not be adequate to keep your new, larger business afloat. Overhead expenses will increase along with your business. These expenses may not be recovered with the pricing structure you are accustomed to using. All in all, growing too fast can be much worse than not growing at all. If you want to expand your company, plan for the expansion. Make financial arrangements in advance, and learn the additional skills you will need to guide the business along its growth path.

Continuing Education

Continuing education is a requirement for all businesses. Many licensed professionals are required by their licensing agencies to participate in continuing education.

In New Hampshire, plumbers must attend an annual seminar before they can renew their licenses. In Maine, real-estate brokers must complete continuing-education courses to keep their licenses active. Whether your business forces you to pursue continuing education or not, you should do so. If you don't keep yourself aware of the changes in your industry, you will become outdated and obsolete. Read, attend seminars, go to classes, do whatever it takes to stay current on the changes affecting your business.

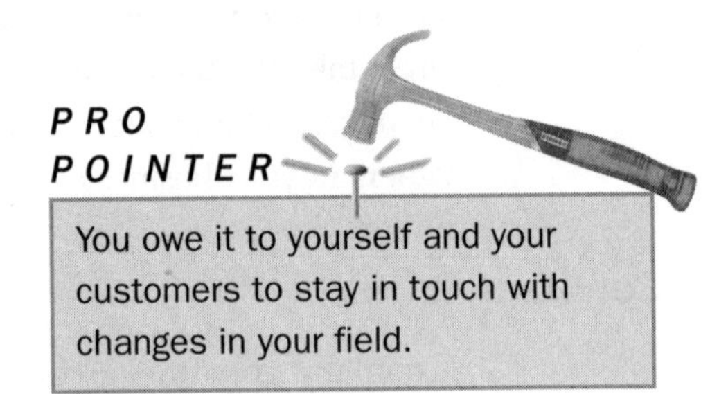

You owe it to yourself and your customers to stay in touch with changes in your field.

Bigger Jobs

Bigger jobs can mean bigger risks. Before you venture into big jobs, make sure you can handle them. There are several factors to consider in your planning. Will you have enough money or credit to keep the big jobs and your regular work running smoothly? Do you have enough help to complete all your jobs in a timely fashion? If you are required to put up a performance bond, can you? Will you be able to survive financially if the money you're anticipating from the big job is slow in coming? Do you have experience in running large jobs, which sometimes from a paperwork standpoint can be time eaters? This line of questioning could go on for pages, but all the questions are viable ones to ask yourself. You shouldn't tackle big jobs until you are sure you can handle them.

Should You Diversify Your Company?

Should you diversify your company? At some point you will probably ask yourself this question. To answer it, you will have to spend some time thinking, evaluating, and researching.

There is no question that diversifying your company can bring you more income. But it can also cause your already successful business to get into trouble. There are many reasons why companies that diversify fail. When you split your interest into multiple fields, you are less likely to do your best at any one job. For this reason, many companies do better when they don't diversify.

In rural areas, it is sometimes necessary for a small business to fulfill many functions to survive. For example, a homebuilder might have to take on remodeling jobs.

In areas with large populations, this type of spreading out is not needed. Builders can build, and remodelers can remodel; there is enough business to go around to all the specialists.

Your geographic location could be a factor in your decision to expand your business operations. The desire to make more money, however, is the most common reason for business alterations. Greed can be a very powerful destroyer. If you try to get your fingers into too many pies, they may all fall off the windowsill.

PRO POINTER

Don't allow yourself to be driven by greed.

If you want to diversify, do so intelligently. Don't just decide one day that you are going to hire a master electrician and expand your building business to include electrical services. What will you do if your master electrician quits? Without a master's license that phase of your business will disappear.

There are many considerations to think about before dividing your time and money into separate business interests. Most people struggle to keep one business healthy. If you get aggressive and open several business ventures, you may find you will lose them all.

Positioning your company to deal with change is an ongoing activity. To maintain your business, you must occasionally stop and look at what you're doing now•what's working, what's not working•and possibly change your plans. Routine adjustments are normal and should be expected. The concepts needed to make the most of your time and money are simple. If you follow the advice in this chapter, you should be able to do very well. Getting too big too quickly is a common cause of business failure, so pace yourself carefully.

CHAPTER SIX

Your Business Structure

Corporations, sole proprietorships, and partnerships are all common forms of business structure. Each form of business has its own set of advantages and disadvantages. This chapter is going to help you explore the different forms of business and decide which one best suits your needs.

You may find that incorporating your business will provide additional security from personal liability. You might also discover that if you incorporate, your tax consequences are worsened. The lure of partnerships may not be so attractive after you read about what can happen to you in a partnership. In any event, by the time you finish this chapter, you should have a clear insight into what type of business structure is best for you.

Choosing Your Business Structure

Choosing your business structure is not a task that should be taken lightly. There are many considerations to evaluate before setting a business structure for your venture. Do you know the difference between a standard corporation and an S-corporation? Are you aware of the tax advantages in operating a sole proprietorship, and are you aware of some pitfalls to avoid? Were you aware that having a partner can ruin your credit rating? If you answered no to these questions, you will be especially glad you read this chapter.

Before we get started, let me say that I am not an expert on tax and legal matters. The information in this chapter is based on experience and research. As with all the information in this book, you should verify the validity of the information before using it. Laws change and different jurisdictions have different rules.

What Is a Corporation?

What is a corporation? A corporation is a legal entity. To be a legal corporation, it must be registered with and approved by the secretary of state. A corporation can live on in perpetuity.

PRO POINTER

The information in this chapter has serious legal and financial consequences, and although I have given you my thoughts on various types of business organizations, you are well advised to seek professional assistance once you decide on the form your business is planning to take.

A Subchapter S corporation is not like a standard corporation. Commonly called an S-corporation, these corporate structures must meet certain requirements. For example, an S-corporation may not have more than 35 stockholders. It is not taxed in the same way as a regular corporation.

Additional requirements include a provision whereby stockholders must submit their individual personal tax returns and show their share of capital gains and ordinary income. Tax preference must also be provided personally by the stockholders. The corporation files a tax return but does not pay income tax. This type of corporation allows the stockholders to avoid double taxation. However, Subchapter S corporations may not receive more than 20 percent of their income from passive income. Examples of passive income are rent from an apartment building, book royalties, earned interest, and so on.

What Is a Partnership?

A partnership is an agreement between two or more people to do business together. In a general partnership, each partner is responsible for all of the partnership's debt. This is an important factor because if you have a bad partner, you could be held liable for all partnership debts. The partnership doesn't pay an income tax, but it must file a tax return. The income from the partnership is distributed to the partners, who include it in their own personal tax returns.

Partnerships can consist of all general partners or one general partner and any number of limited partners. General partners do not have limited liability. The general partner is accountable for all actions of the partnership. Limited partners can have limited risk. For example, a limited partner may invest $10,000 in the partnership and limit his or her liability to that $10,000. Then, if the partnership loses money, goes bankrupt, or has other problems, the most that the limited partner stands to lose is the investment of $10,000. The general partner, however, will be held responsible personally for all actions of the partnership.

What is a Sole Proprietorship?

A sole proprietorship is a business owned by an individual. The business must file a tax return, but any income tax is assessed against the individual's tax return. Sole proprietors are exposed to full liability for their business actions; however, they can purchase various types of insurance to limit their liability. Now you know what each type of business structure is, so let's see what type fits your needs.

What Type of Business Structure Is Best for You?

Now that you know what the different types of business structures are, which would you choose if you had to decide now? I haven't given you enough information to make a wise decision yet. Before you make a firm decision, you owe it to yourself to read and digest the remainder of this chapter. It will be interesting to see if your choice after reading the entire chapter is the same as it is now.

Corporations

The primary advantage of a corporation for most business owners is the limited-liability aspect, but the protection may not be as good as you think. Many business owners don't know how corporations work. These same owners will pay the fees to incorporate without realizing that they are not getting what they are paying for. They are getting a corporation, but they are not buying the protection they think it involves. There are several misconceptions about corporations and the limited-liability issue. Let's explore some of them.

What a corporation may or may not protect you from depends on what type of liability you are seeking protection from and how your business operates. Let me be more specific on these issues.

PRO POINTER

Protection from lawsuits is one consideration when deciding to form a corporation. It is almost impossible to protect yourself entirely from lawsuits, but a corporation can help. If your business is a corporation, your liability may be limited to the corporate assets, but don't count on it.

Assume that your business is a corporation. You are a remodeler and your own lead carpenter. When you have a house to remodel, you take an active part in the construction as a carpenter. This means that you are nailing nails and doing much of the physical work involved with the construction of each home built. On one particular job, you've cut a hole in the subfloor to allow your mechanical trades easy access to the basement. Stairs to the lower level have not yet been installed. Workers are moving up and down with the use of ladders. A piece of plywood is nailed in place over the opening at the end of each day. However, one day you forget to nail the cover in place. Your customer comes out to inspect the day's work, falls through the hole in the subfloor, and decides to sue. Can the customer sue the corporation and go after its assets? Yes, the customer can sue the corporation. Can the customer sue you as the primary stockholder of the corporation? Not really. I suppose anyone can sue anyone else for any reason, but it would be unlikely that the customer would get far in a lawsuit against you as a stockholder. But, don't relax. You're not safe. Since you did the work yourself, the customer can sue you individually for the mishap. This is a fact that most business owners don't realize. If you had sent an employee or a subcontractor out to do the work, your risk of being sued for your personal assets other than your corporation would be minimal.

PRO POINTER

If you don't do your own fieldwork, a corporation offers some protection for your personal assets.

Personal financial protection is usually another goal for people who want to incorporate their business. Some of these people feel they can limit their financial liability with the corporation, and to some extent this is true. If a corporation gets in financial trouble, it can declare bankruptcy without the loss of personal assets belonging to stockholders. However, most lenders and businesses that extend credit to small corporations often require someone to sign personally for the debt. Before you decide to incorporate, talk to experts for details on how incorporating will affect you and your business.

Partners and Partnerships

PRO POINTER

If you endorse a corporate loan personally, you will be responsible for the loan, even if the corporation goes into bankruptcy.

Dealing with partners and partnerships can get tricky. My worst business experiences have involved partners. It would be easy to think that my bad track record with partners has been my fault, but the clients whom I consult with have shared similar bad experiences with partners. I can think of very few successful partnerships. If you are considering going into business with a partner, go into the arrangement with your eyes wide open.

If you are a general partner, you can be held accountable for the business actions of your partner or partners; however, once again, there are various types of insurance that can provide some limited protection. This fact alone is enough to make partnerships a questionable option.

If you decide you want to set up a partnership, consult an attorney. The money you spend in legal fees forming the partnership will be well spent. It is much better to invest in a lawyer to form the partnership than it is to pay legal fees later to resolve partnership problems. Now, let's look at the pros and cons of each type of business structure.

PRO POINTER

There are very few advantages to partnerships, but the potential problems are numerous.

Learn the Pros and Cons of Each Form of Business

Before you can make a final decision on what type of business structure to assume, you should learn the pros and cons of each form of business. Let's take each type of business structure and examine the advantages and disadvantages.

The Advantages and Disadvantages of Forming a Corporation

Corporate advantages are abundant for large businesses, but a standard corporation may not be a wise decision for a small business. What advantages does a standard corporation offer? Depending on the size and operational aspects of a business, a corporation can provide protection from personal lawsuits and financial problems.

Corporations issue stock of several types, two of which are "voting" and "non-voting" stock, which as their name implies allows the stockholder to vote on company business or have no legal voice in corporate affairs. Offering non-voting stock to attract an exceptional employee to join the company is often used as an incentive to join the company. Since corporations can issue stock, there is the possible advantage of generating cash from the sale of stock.

While small businesses can suffer from double taxation with a standard corporation, they can benefit from some tax advantages. By incorporating your business, some expenses that are not deductible as a non-corporate entity become deductible. Insurance benefits could be an example of this type of deductible advantage.

In many ways the corporate disadvantages are more significant than the advantages for a small business. The first disadvantage is the cost of incorporating. There are filing fees that must be paid, and most people use lawyers to set up corporations. The cost of establishing a corporate entity can range from less than $200 to upwards of a $1,000. This can be a lot of money for a fledgling business.

There is more to a corporation than setting it up and forgetting about it. To keep the advantages of a corporation effective, you must maintain certain criteria. You will have to have a registered agent. Many people use their attorney as a registered agent. If you use an attorney in this position, you will be spending extra money. There are board meetings to be held, and the corporate book must be maintained. A board of directors must be established. Written minutes must be recorded at these meetings. Annual reports are another responsibility of corporations. Again, many people have an attorney tend to much of this work.

Small business owners who incorporate their business with a standard corporation must face double taxation. These owners will pay personal income tax and will also pay corporate taxes. This extra taxation is a serious burden for most small businesses.

PRO POINTER

Corporate officers must be appointed, and the fees for maintaining an effective corporation can add up. If the corporate rules are broken, the corporation loses much of its protection potential. If an aggressor can pierce the veil of your corporation, you will have personal liability.

The advantages of a Subchapter S corporation are a little better than those of a standard corporation for a small business. One of the biggest advantages of a Subchapter S corporation over a standard corporation is the elimination of the double taxation feature. Stockholders of Subchapter S corporations pay personal taxes on the money they and the corporation earn, but they pay one tax on the money earned. S corporations offer the other advantages, such as limited liability, of standard corporations.

The disadvantages of S corporations are not too bad. The costs of setting up and maintaining the corporation are still there. If you plan to have more than 35 stockholders, you will not be able to use an S corporation. If more than 20 percent of your corporate income will be passive income, you cannot use a Subchapter S corporation.

The Advantages and Disadvantages of a Partnership

The advantages of a partnership are not too numerous. The cost of establishing a partnership is less than incorporating, but I don't see the need for partnerships in small business operations. Partnerships have their place in business ventures, but I don't like them for most types of business. If you are a real-estate investor, partnerships can work, but why gamble with a partnership for a service business? Set up a corporation or investment agreement but avoid partnerships.

In my opinion, the disadvantages of partnerships are abundant. When you are in a partnership as a general partner, you share a portion of the profits and all the risk. Hooking up with a bad partner can not only ruin your business; it can ruin your future. Some people prefer business partnerships, but I'm not one of them. If you decide to set up a partnership, get legal counsel to understand what you might be getting into.

The Advantages and Disadvantages of a Sole Proprietorship

The advantages of a sole proprietorship are many. A sole proprietorship is simple and inexpensive to establish. You are your business, so there are no complicated corporate records to keep. You can get tax advantages as a sole proprietor, as you can with other forms of business structure. You are the boss; there is no partner to argue with over business decisions. There are no stockholders to report to. Tax filing is relatively simple, and you don't have to share your profits with others.

PRO POINTER

You might take a sheet of paper and divide it down the middle—make the left side Advantages and the right side Disadvantages and then list each type of business entity you are contemplating. Sometimes a visual presentation will help you make up your mind. Take the time to make a wise decision before you cast the structure of your business in stone.

There are some disadvantages to a sole proprietorship. As a sole proprietor, you will have to sign for all your business credit personally. If your business gets sued, you get sued. There may be some tax angles that you will miss out on as a sole proprietor.

There is no easy answer for which type of business structure is best for you. Circumstances dictate the ideal type of organization to form. Talking with your attorney and your accountant is a good idea when you are trying to decide which path to take in establishing your business identity.

CHAPTER SEVEN

Office and Storage Requirements

Office space can be a major factor in the successful operation of your business. There is little need for a construction business to have an office in the high-rent district. Most contracting businesses can function from a low-profile location. This is not to say that office space is not needed or is not important. Whether you work from your home or a penthouse suite, your office has to be functional and efficient if you want to make more money.

Should You Work from Home or from a Rented Space?

This can be a very difficult question to answer, but with some thought and based upon some of my experiences and advice, you can make an intelligent decision. I have worked from home and from commercial offices, and my experience has shown that the decision to rent commercial space is dictated by your own self-discipline, the type of business you are running, and cost. Let's explore the factors you should consider when thinking about where to set up shop.

Self-Discipline

Self discipline is paramount to your success in business, and the ability to work out of an office in your home is a good test. It is easy to spend too much time around the breakfast table or go for a stroll around your farm. Working from home is very enjoyable, but you do have to set rules and stick to them. Designating a specific room for your home office, hopefully away from daily household activity, is a start in the right direction.

Storefront Requirements

Some businesses require storefront exposure, but remodelers rarely have such needs. Other trades, such as plumbers or electricians, may need to display the fixtures they sell, but most remodelers have no need to do this. They can send their customers to supplier's showrooms or show them catalogs of plumbing and electrical fixtures, kitchen cabinets, windows, doors, and hardware selections. Having an office that your customers can visit helps to build professionalism, and this is an important element for success. But your first step is looking at your finances•can you afford the expense of a rented office?

Home Office

A home office is a dream of many people. Putting your office in your home is a good way to save money, if it doesn't cost you more than you save. Home offices can have a detrimental effect on your business. Some people will assume that if you work from home you are not well established and may be a risky choice as a contractor. Of course, working from home doesn't mean your business is having financial trouble, but some customers are not comfortable with a company that doesn't have a commercial office space.

I work from home now and I have worked from home at different times for nearly twenty years. I love it. I am also very disciplined in my work ethic, and even though my office is in the home, it is set in professional style. When clients come to my home office, it is obvious that I am a professional. I will talk more about setting up a home office a little later, but take your home office seriously. Having an office in your home allows you to write off certain household expenses. By calculating the size of your office as a percentage of the total square footage of your office, you may be able to deduct a certain portion of utility costs, real-estate taxes, and mortgage payments .All costs to furnish your office are generally tax-deductible as well.

PRO POINTER

If you are considering setting up an office in your home, you ought to talk to an accountant and find out the tax advantages and tax rules pertaining to a home office.

Commercial Image

A commercial office can give you a commercial image. This image can do a world of good for your business. However, commercial office space can be expensive and add greatly to your overhead expense. Before you jump into an expensive office suite, consider all aspects of your decision. We will talk more about the pros and cons of commercial offices as we continue with the chapter.

Assess Your Office Needs

Before you can decide on where to put your office, you need to assess your office needs. This part of your business planning is important to the success of your business. Can you imagine opening your business in an expensive office and then, say six months later, having to move out because these expenses are hurting you? Not only would that situation be potentially embarrassing, it would be bad for business. Once people get to know office location, they expect it to stay there or to move up. A downward move, like the scenario described, would concern certain people who may think your business is not doing well and scare off a percentage of future customers.

How Much Space Do You Need?

One of the first considerations in choosing an office is how much space you need. If you are the only person in the business, you may not need a lot of space in the office. When you consider your space requirements, take the time to sketch out your proposed office space. It helps if you make the drawing to scale. You can also make some templates, to scale, of a desk, chair, filing cabinets, storage shelves, and a table for a copier. Move them around in the space you think you need for an office to determine whether you need more or less than you originally thought. How many people will you meet with at any given time? How many desks will be in the office? I'm used to be a one-man business, not counting subcontractors, but I had two desks and a sorting table in my primary office. In addition, I had another room designated as my library and meeting room, another space set up as a darkroom, and a photography studio in my basement. My barn stored my tools, equipment, and supplies. So, you see, even a small business can have the need for large spaces.

When you are designing your office, consider all your needs. Desks and chairs are only the beginning. Will you have a separate computer workstation? Do you need a conference table? Where will your filing cabinets go? Where and how will you store your office supplies? How many electrical outlets will be needed? The more questions you ask and answer before making an office commitment, the better your chances are of making a good decision.

Do You Need Commercial Visibility?

PRO POINTER

Remember to leave enough room on the scaled drawing to allow your chair to be pushed back from your desk and not hit any other furniture.

Do you need commercial visibility? Almost any business can do better with commercial visibility, but the benefits of this visibility may not warrant the extra cost. If you are out in the field working every day and you don't have an employee in the office, what good will it do you to have storefront exposure? If your business allows you to remain in the office most of the time, a storefront might be beneficial. You would get some walk-in business that you wouldn't get working out of your home.

Do You Need Warehouse Space?

If you have to keep large quantities of supplies on hand or deal in bulky items, warehouse storage may be essential. One solution for office and storage space is a unit that combines both under the same roof. These office/warehouse spaces are efficient, professional, and normally not very expensive if you can find them.

If you don't need access to the materials in storage, renting a space at a private storage facility can be the best financial solution to your needs. At one time, I ran my company from a small office and a private storage facility. This arrangement was not convenient, but it was cost- effective and it worked. Once you have evaluated your needs, you can consider where to put your office.

Location Can Make a Difference

In business, location can make a difference. If you cater to customers in the city, living in the country can be inconvenient and cause you to lose customers. Having an office where you are allowed to erect a large sign is excellent advertising and builds name recognition. If your office is in a remote section of the city, people may find it difficult to get to your office and stay away. If you work from your home and your home happens to be out in the country, customers may not be able to find you even if they are willing to try. Location is an important business decision.

How will your office location affect your public image? Public opinion is fickle, and if your business is perceived to be successful, it probably will become successful. On the other hand, if the public sees your business as a loser, look out. It is unfortunate that we have to make some decisions and take some actions just to create a public image, but there are times when we must.

Aside from the prestige of an office location, you must consider the convenience of your customers. If your office is at the top of six flights of steps with no elevators available, people may not want to do business with you. If there is not adequate parking in the immediate vicinity of your office, you could lose potential customers. All these factors play a part in your public image and success.

PRO POINTER

What does the location of your office have to do with the quality of your business? It probably has nothing to do with it, but the public thinks it does. For this reason, you must direct your efforts to the people you hope will become your customers.

How Much Office Can You Afford?

This would appear to be a simple question, but answering it may be more difficult than you think. When you look at your budget for office expenses, you must consider all the costs related to the office. These costs might include heat, electricity, cleaning, parking, snow removal, and other similar expenses. These incidental expenses could add up to more than the cost to rent the office.

If you rent an office in the summer, you might not think to ask about heating expenses. In Maine, the cost of heating an office can easily exceed the monthly rent on the space. When you prepare your office budget, take all related expenses into account and arrive at a budget number you are comfortable with and that is realistic.

When you begin shopping for an office outside your home, ask questions and lots of them. Who pays for trash removal? Who pays the water and sewer bill? Who pays the taxes on the building? These questions are important because some leases require you, the lessee, to pay the property taxes. Who pays the heating expenses? Who pays for electricity? Who pays for cleaning the office? Does the receptionist in the lower level of the building cost extra? If there is office equipment in a common area, such as a copier, for example, and if so, what does it cost to use the equipment?

Ask all the questions and get answers. If you are required to pay for expenses such as heat or electricity, ask to see the bills for the last year. These bills will give you an idea of what your additional office expenses will be.

Before you rent an office, consider ups and downs in your business cycle. If you are in a business that drops off in the winter, will you still be able to afford the office? Do you have to sign a long-term lease, or will you be on a month-to-month basis? It usually costs more to be on a month-to-month basis, but for a new business it might be well worth the extra cost. If you sign a long lease and default on it, your credit rating can be tarnished. What happens if your business booms and you need to add office help? If you are in a tiny office with a long-term lease, you've got a problem. If you do opt for a long-term lease, negotiate for a sublease clause that will allow you to rent the office to someone else if you have to move.

Many new business owners may look for space with marble columns, fancy floors, wet bars, and all the glitter depicted in offices on television. Well, unless you are independently wealthy, these high-priced work spaces can rob you of profits and put you out of business. Potential customers may view a fancy office as an expense that is tacked on to the price of the house you are selling. Don't assume that an expensive office is going to return higher net earnings.

PRO POINTER

It can be easy to dream about how a new office will bring you more business, but don't put yourself in a trap. When you project your office budget, base your forecast on your present workload. Better yet, if you've been in business awhile, base the projections on your worst quarter for the last year. If you can afford the office space in the bad times, sign the lease. If you can only afford it during the summer boom, you're probably better off without the office.

Answering Services Compared to Answering Machines

When answering services are compared to answering machines, you may find many different opinions. Most people prefer to talk a live person rather than an electronic device. However, as our lives become more automated, the public is slowly accepting the use of electronic message storage and retrieval.

When you shop for services, which do you prefer, an answering service or an answering machine? Do most of your competitors use machines or people to answer their phones? This is easy to research. Just call your competitors and see how the phone is answered. The use of an answering machine may cause you to lose business, but that doesn't mean necessarily that you should not consider it.

Answering machines are relatively inexpensive, and most machines are dependable. These two points give the answering machine an advantage over an answering service. Answering services are not cheap, and they are not always dependable. Answering services can page you to give you important and time-sensitive messages; answering machines can't. Another point for answering services is that you can access them from your mobile phone and pick up your messages while out of the office.

To determine which type of phone answering method you should use, make a list of the advantages and disadvantages of each. Once you have your list, arrive at a first impression on which option you should choose. It may be necessary to change your decision later, but at least you will be off to a reasonable start.

There are several qualities you should look for in an answering machine. They should include the ability to check your messages remotely. Most modern answering machines can be checked for messages from any phone. Choose a machine that allows the caller to leave a long message. Many machines will allow the callers to talk for as long as they want. These machines are voice-activated and will cut off only when the caller stops speaking. Pick a machine that will allow you to record and use a personal outgoing message. Some of the answering machines are set up with a standard message that you can't change. It will be beneficial to customize your outgoing message. If you buy an answering machine that meets all these criteria, you should be satisfied with its performance.

In considering answering services, price is always a factor, but don't be guided solely by price. You get what you pay for. Find an answering service that answers the phone and takes messages in a professional manner. You want a dependable service, one that will deliver messages to you in a timely manner.. Ask if you can provide a script for the operators to use when answering your phone. Some services answer all the phones with the same greeting, but many will answer with your personalized message.

Ask about the hours of the day for your coverage. Most services provide 24-hour service, but that generally costs more. Ask if the service will page you for time-sensitive calls; most will. Determine if your bill will be a flat rate or if it will fluctuate, depending on the number of calls you get. Inquire about the length of time you must commit to the service. Some answering services will allow you to go on a month-to-month basis, and others want a long-term commitment.

If you decide to use an answering service, check on the performance periodically. Most services provide a special number for you to call to pick up your messages, especially if they base your bill on the number of calls received. Even if you will have to pay for calling in on your own line, do it every now and then. When you call in on your business number, you can get first-hand proof of how the operators handle your calls. Have friends call and leave messages. The operators won't recognize the voices of these people and will treat them like any other customer. This is the best way to check the performance of human answering services.

> ***PRO POINTER***
>
> Machines or humans? I think the business you lose with an answering machine is more valuable than the money you save. If you can hire a personalized answering service, you should. I have tried having my phones answered each way, and I am convinced that human answering services are the best way to go.

Pulling together all the components of a functional office can take time. Don't expect to make the best decisions on your first attempt. You may have to experiment to find out what works and what doesn't. Remember this, don't jump into a long lease. You are better off growing slowly than not growing at all.

CHAPTER EIGHT

Adding a Computer to Your Toolbox

A computer can be a valuable tool in building your business in so many ways–from producing professional-looking letters to assisting you in the design and estimating of your jobs. Bucking the trend of most other products, the cost of computers and their peripheral equipment has plunged over the years. In the 1980s a high-quality computer could cost you $2500 and a laser printer another $650. But today, a top-quality desktop computer with triple the memory power of those 1980s jobs can be purchased for under $450. By adding $125 for a really good printer you can now have enough equipment to perform lots of tasks. Upgrading may only become necessary when you decide to add computer-assisted design (CAD) software, which may require increased hard-drive capacity and speed.

By adding Internet service, a whole new world opens up. There are construction web sites such as McGraw Hill Construction and the National Association of Home Builders (NAHB) that provide information about industry trends, statistical data, research, and educational material. Equipment and product manufacturer's web sites can be tapped for detailed information about their products.

PRO POINTER

Software available on the market today allows you to computerize your estimating or set up your own bookkeeping system—with step-by-step instructions on handling accounts receivable, payables, customer invoicing, and check processing.

There are many trade associations such as the Southern Pine Council (softwoods), Western Wood Products Association (framing lumber), Brick Institute of America, and National Wood Window and Door Association that maintain web sites for the purpose of disseminating technical product information–much available at no cost, some at nominal cost. So a computer really becomes your window on the world of construction.

Desktop Alternatives

Portability is the latest development in the computer field and laptops today are thinner and lighter with bigger screens and more hard-drive memory that ever before. As of this writing you can buy a laptop with a 60-gigabyte hard drive, 15-inch screen, and mobile technology connectivity for about $1,500. A top-notch brand with "bare bones" features and a 60-gig hard drive can be purchased for about $1,000.

Notebooks, a size smaller than laptops, with a swiveled 12-inch screen can be a very useful tool for builders . The software allows for a lined "notebook" page to appear on the screen, and it will accept both handwritten notes or hand-drawn sketches. With wireless capability, these notes or drawings can be instantaneously sent to a millwork company to show what you need to be priced, or a sketch can be sent to your customer's computer to confirm a requested change. While working around the construction site, you can jot down notes, make changes in some dimensions, or plan your next day and upon return to the office either download into your desktop or print out this information.

Another step in portability and convenience is the personal digital assistant (PDA), a pocket-sized device that used to hold telephone numbers and your appointment schedules. Increased power and special software programs just for builders make these powerhouses more useful than ever. Newer models allow for hands-free operation and respond to voice commands so that you can dictate while you drive and download when you return to your office. There are punch-list and inspection programs available, and undoubtedly more will come on the market in the weeks and months ahead. So you have a wide range of computer technology available–choose wisely and buy only what you need

Getting Started

I used to hate computers. One of my workers went about her business with the computer trying unsuccessfully to get me interested, but when she wanted to put my business on the computer I told her that would never happen. Well, was I wrong.

She began to use the computer for some business and office applications, and I immediately saw that many of these functions were being handled more quickly and more accurately. At first, my assistant handled all the computer operations, but in time I started to play with the system. At first, I was often frustrated, but it didn't take me long to see how this technology could be a tremendous help in my business. It wasn't long before we upgraded our system to a more powerful one and the rest is history. I now know I rely on my computer for a multitude of tasks.

It starts with using a word-processing software package. You see how much easier it is to prepare your contracts and correspondence on the computer if you grew up with the typewriter. The ability to create forms and store them on a disk or on your computer's hard drive makes it fast and easy to turn out proposals, letters, and much more.

The next step is spreadsheet software. As the name implies, you can create an electronic spreadsheet with budget numbers or estimate breakdowns. By changing one or two numbers, the total is automatically changed to reflect the increase or decrease. Suddenly, forecasting future income and expenses and tracking your budget becomes fun. You are starting to get hooked.

Then you experiment with the database programs. Mailing lists were never so easy to accumulate and use. Your marketing will be much easier with your new, computerized customer base. Checking inventory is a snap, and storing historical data is a breeze.

With a little more time, you get into accounting programs and automated payroll. With a little playful study, you find you can do what you've been paying other people to do for you. For the one-time cost of your software, you have eliminated a routine overhead expense.

PRO POINTER

Once you go online, you're downloading bid-sheet information, scanning job opportunities, and exchanging comments with other computer users. It doesn't take long for the appeal of computerization to sink in.

How Will a Computer Help Your Business?

A computer can help your business in a myriad of ways. You can perform most clerical or financial functions on a computer. You can draw blueprints and various types of house plans with CAD software. Marketing and advertising will be easier, and you can create your flyer or ad on the computer, change elements around until you have the right format, and if you wish, print out one-page flyers that have a professional look.. Inventory control and customer billing will almost take care of themselves on a computer. As you explore the various programs, you will see that there are dozens of ways for your business to benefit, so let's take a closer look at some of them.

Marketing

Marketing can be done very efficiently with a computer. You can build a database file of your customers and potential customers. Then you can print mailing labels and send out flyers and announcements to your current and potential customers with ease. If you use telemarketing, you can use computers to conduct your cold-calling surveys.

Artwork software enables you to create logos and flashy flyers. By using the creative options available through your computer and drawing program, you can attract more attention and reduce your outside printing costs.

Tracking historical data is easy with a computer. With just a few keystrokes you can see, for example, what decks were selling for two years ago. A few more taps on the key can display the type of work that produced the most profit for you over the past months or year.

Payroll

By entering a minimal amount of data, the computer will run your payroll for you. This is a real time-saver for companies with several employees. As you get to be more comfortable and more knowledgeable about computer operations, developing your own custom software programs can turn a payroll register into an estimating tool. When you produce your weekly payroll checks for your field workers, you need to know the number of hours they worked and their hourly rate of pay—but by adding one more bit of information, you can also produce unit costs for each operation being performed during the pay period. You first need to identify the task that was performed for the hours reported—say, for sheetrocking. By asking the worker to report the total square footage of sheetrock installed during that pay period, it is a simple matter to divide this amount by the total labor cost reported in the payroll register, and you now have the cost per

PRO POINTER

If you have employees on the payroll, you will enjoy the automated features available in payroll software.

square foot of sheetrock installed. When this exercise is repeated whenever sheetrock is installed, by averaging all of the unit costs you can get a pretty good idea of how much it costs to install drywall. This same process can be used to develop costs for concrete work, framing, painting, flooring, and any other job.

Job-Costing Results

When you have completed your job and have all of the bills, don't just file them away—they hold a wealth of information. For example, calculate the type of work that you have completed. How much was the total cost? Keep records of your costs to learn from.

You can go even further if you like. How much did it cost to install that hardwood floor in the living and dining room? By keeping track of those labor costs during construction and adding material and finishing costs, then dividing by the installed area, you now have the cost per square foot of hardwood flooring. If you set up a series of historical cost files in your computer, you can enter the cost for each hardwood floor, date installed (even break down costs into labor and materials), and create a meaningful database of your costs.

PRO POINTER

Job-costing results can be printed out of your computer files in minutes. Not only will computer-generated job-costing numbers come more quickly, but they will probably be more accurate.

Tracking Your Budget

Tracking your budget is relatively easy when you use the right software. You can boot up the computer and see where you stand at any time during the year. This is not only convenient but you can track costs versus budget periodically.

Projecting Tax Liabilities

Projecting tax liabilities is another function a computer can perform for you. If you are wondering how much money you will need at tax time, just refer to the information stored in your computer. Without hesitation, the computer will call up the information you need to plan for your tax deposits.

Computerized Estimating

Computerized estimating will make your life easier. There are systems available that make estimating an easier task. One such system uses a "light pencil" to count items and a database containing labor and material costs. Using doors, frames, and hardware as an example, you create your own database by punching in your costs to, say, purchase a wood door/frame and hardware. You also punch in the labor cost to install this pre-hung door. By tapping the light pencil on each door of that type on the plans, the computer calculates the cost per door, adds up all the doors, and gives you a total cost for all doors of this size. Using the light pencil to trace the outline of the kitchen floor, for example, when you have previously punched in the cost per yard of resilient flooring, the computer will calculate the cost of the flooring. Not only will your estimates be more accurate, but you will be able to prepare them much more quickly.

Word Processing

Word processing will make all your written work easier. With built-in spellcheckers and thesauruses, today's word-processing software takes the drudgery out of writing a letter. If your grammar skills are rough around the edges, you can incorporate grammar software to correct your writing.

Databases

Databases can be used to store information on any subject. Whether you want to know the cost of a ton of topsoil or when your trucks are due for service, a database can do the job. You can design the database file to include as much or as little information as you like. Sorting these electronic files is much easier than digging through the old filing cabinet.

Checks and Balances

If you don't like writing checks or keeping track of your bank balance, let the computer do it for you. There are dozens of programs and forms available for paying your bills with a computer. As a side benefit, the computer will make adjustments to your bookkeeping records as it goes.

Customer Service

Customer service can be improved with the use of computers. If you receive a warranty call, you can quickly determine if the job is still covered under its warranty. When your customers are having a birthday, your computer can remind you to send out a birthday card. Whatever reason you have for wanting to stay in touch with your customers, a computer will allow you to do it faster and easier.

PRO POINTER

By keeping track of all of your warranty items, you can review them from time to time to find out where the most frequent warranty and maintenance items occur. Then you can take steps to reduce them, adding to your customer good will.

Inventory Control

Inventory control is made faster and more efficient with the use of computers. When you punch in the minimum quantity of a certain material or item you must have in stock and insert all withdrawals from that stock, you computer will flag the need to replenish. At the end of the year, you won't have to spend hours going into the back room, counting your inventory. All you have to do is have the computer print a report, and in moments your inventory is done.

Personal Secretary

With a memory-resident program, your computer becomes a personal secretary. The machine will beep to remind you of your next appointment. You can store phone numbers in the computer and have it dial calls. You can even send a fax. The right software will all but replace your old appointment book, and with the touch of one key, you can scan all your appointments as far in advance as you like.

Building Customer Credibility with Computers

Computers can lend an air of distinction to your company. Customers can contact you via email, and you can reply by sending a proposal, a sketch of a house plan, or, via a digital camera, even pictures of one of your remodeling jobs. It is fairly easy and not too expensive to create your own web site and invite potential buyers to tap into it.

Your company logo, current projects, some interior shots of recently completed jobs, even a testimonial from a couple of satisfied clients will not only build your credibility but inform interested buyers about your company and your product.

Direct-Contact Approach

The direct-contact approach is one that involves inviting customers to come into your office and observe your computer operation, your computer assisted design capability, your ability to track costs for your billings, and even to show them some portions of your database and how you develop estimates. There is no doubt that the visual effect of the hardware, your software programs, and your ability to use them will impress the customer.

Let's create a hypothetical situation to elaborate on the use of your computer. A young couple comes to your office to discuss plans that they have for a major room addition, and they have a rough sketch of what they want, but it is not drawn to scale. Other remodelers they have visited have merely made photocopies of the sketch and let it go at that. But you are going to be different. You are going to make a lasting impression on this young couple by demonstrating your sophisticated electronic equipment and your ability to use it.

You take their sketch, put it in your flatbed scanner, and transfer it electronically to your computer. The young couple suddenly sees the drawings on your computer screen—but they certainly are surprised when they see that you have taken this rough sketch and converted it into a scale drawing. Using your CAD program (and you explain to them what it's all about) you can begin to customize the drawing. Do you think you'd like the bedroom on the other side of the bathroom? How about a walk-in closet over here? And the closet appears immediately. Kitchen a little too small? Let's increase the width by 1 foot—and so on.

As you talk to the potential customers, you make adjustments in the on-screen drawing, moving the kitchen sink down the counter, adding an island workspace, drawing in a fireplace, and so on. In less than an hour, you have a viable plan drawn for the anxious buyers.

Now tell me, which contractor would you hire to build your house, the one who made a photocopy or the one that took the time and had the technology to produce a professional drawing? I think the answer is obvious. Even if the contractor with the ability to do all this with the computer wanted a little bit more money for the same job, the prospective customer would probably assume he or she was worth it. It is not too difficult to see how your computer can become a powerful marketing and sales tool.

The Indirect Approach

The indirect approach can be equally effective. Assume for a moment that you are a prospective homeowner. You are soliciting bids for a new sunroom. During your search, you have narrowed the field of general contractors down to two. However, the two contractors are evenly matched in their prices, references, and apparent knowledge of building. Who will you choose? You must look deeper to find what separates these two contractors. You start your reconsideration by going back over the bid packages.

The first contractor, Harry's Home Remodeling, Inc., has submitted a bid package similar to the other general contractors. You received a fill-in-the-blank proposal form, the type available in office-supply stores. The proposal had been typed, but it was obvious that the preparer was not a typist. Harry's references checked out, even if they were handwritten on a piece of yellow legal paper. The plans submitted were drawn in pencil. The graph paper was stained, probably from coffee. While there was a spec sheet in the package, it was vague. In general, the bid package was complete, but it wasn't very professional.

Harry's competitor, Cornerstone Builders, Ltd., submitted essentially the same information, but the method of presentation was considerably different. The contract from Cornerstone was clean and neatly printed with a laser printer. The paper was heavy, high-quality stock. The accompanying specification sheet was very thorough. It listed every item and specified the items in complete detail.

The reference list supplied by Cornerstone was printed with the same quality as the contract, and the references were all satisfied customers. The plans in this bid package were clean and precise, created on the computer, and contained floor plans, interior and exterior elevations, a cross-section through the building, and even an enlarged view of the kitchen or bathroom. The bid package from Cornerstone Builders, Ltd., was inserted in an attractive binder and spoke professionalism. In this scenario which contractor would you choose? I believe you would have picked Cornerstone. A good presentation can get the job, and a computer can help you make really great presentations.

Spreadsheets, Databases, and Word Processing

Spreadsheets, databases, and word-processing software applications are what most business owners use. Each of these programs can be purchased as stand-alone software, or you can buy a software package that incorporates all these features. These combination packages are called integrated packages. Both types of software can be good, and what you choose will depend on your needs and desires. Let's take a look at some of the pros and cons of each type of software.

PRO POINTER

If you are new to computers, an integrated package may be your best choice. Many stand-alone programs are more complex than the individual modules in integrated packages.

Spreadsheets

Spreadsheets serve many business functions. They allow you to create columns for accounting, estimating, or presentation purposes. Spreadsheet software allows you to insert rows and columns of numbers. When one number is changed, the remaining numbers or column totals automatically adjust to reflect this change. Choosing the best spreadsheet for your business needs may require some research. There are dozens of programs that can get the job done. Some of these programs cost hundreds of dollars, and others can be had for about $50.

You don't have to spend a fortune for software. There are many good deals available—just check with any electronic or office-supply company in your area or go online. Some of this bargain-basement software is great, and some of it leaves a lot to be desired. However, for less than $5, you can often "test drive" the software by buying a demo disk. By shopping around, you can find lots of software combinations. If you opt for major-brand software, you may need some time to get acquainted with your purchase.

PRO POINTER

If you buy brand-name software, there are some excellent tutorial books available to help you learn the software applications. There are also some tutorial CDs that walk you through the instructions by creating computer simulations on your screen.

Most spreadsheets will perform approximately the same functions. Some are easier to learn than others, but the end result will be about the same. When you are ready to buy a spreadsheet program, do your homework. There are many good programs available.

Databases

A database is nothing more than an electronic file, but opposed to that folder in your filing cabinet, this file can be activated by a keystroke. This electronic file cannot only be accessed in the wink of an eye, but it can be changed—updated, erased, added to just as quickly. The storage capacity of these electronic files boggles the imagination—how about six billion bits of data? An entire book can be stored on one 3½-inch diskette, and even more data can be stored on a compact disk. Printing reports and mailing labels are no problem when you have a database program. If you would keep it in a filing cabinet, you can keep it in a database.

PRO POINTER

Even with the storage capacity in budget-priced computers today, their hard drives can provide you with nearly unlimited chances to store and retrieve information. Sorting and retrieving information in a database is easy.

Word Processing

Word processing is probably one of the most popular software programs available today. This type of software is great! You can write and store form letters. You can merge names into the form letters to achieve a personal mailing. If you make a mistake in your typing, you don't need correction tape or correction fluid. You simply type over your mistakes. Word-processing programs can be used for all your correspondence needs, as well as the creation of custom forms, and with the addition of artwork software, it is a terrific way to produce eye-catching promotional material.

Nowadays computers come preloaded with word-processing software, and it seems like the only decision you have to make is whether you want to purchase voice-recognition software that allows you to speak into the computer, which will then convert speech patterns to the written word.

And you don't even have to worry about spelling, or in some cases, sentence structure. Most word-processing software programs today include spellchecking. You are alerted to the misspelling of a word when the computer underlines that word in red. Simply clicking on the spelling icon will give you the correct spelling. When your sentence structure doesn't seem quite right, the computer will underline that sentence in green, allowing you to review it and make changes. Other software programs that are frequently included in the word-processing package include a thesaurus—actually a dictionary of words having a similar meaning.

Integrated Programs

Integrated programs can fill all your basic needs. These programs include spreadsheet, database, and word-processing programs in a single piece of software. These programs are efficient and relatively inexpensive to purchase if they haven't been already incorporated in your new computer. Some software manufacturers offer upgrades when you purchase a computer. Microsoft has an Office package that contains many additional business functions. There are several construction-estimating programs that also include an accounting package that in effect takes your estimate and coverts it into a job-cost reporting system and sets up accounts payable and receivables.

PRO POINTER

Some of these integrated software programs run into hundreds of dollars, so check them out thoroughly before purchasing them—you may find that you have bought a system that is meant for a multimillion-dollar contractor, most parts of which you'll never use.

The Power of Computer-Aided Design (CAD)

If you design your own home plans, you need to learn how computer-aided-design (CAD) programs can change the way you do business. CAD programs allow users to create professional-looking plans even though they have not studied architecture or have little ability to draw. I'm a good example of the type of person that benefited from a CAD program.

My first attempt at CAD was a miserable disappointment, but my further attempts produced excellent results. After I got into computer drafting, I saw a way to improve my contracting business. I gave you an example earlier about how a contractor captures business with a CAD system. This procedure has proved successful for me.

The results you can achieve with a computer-drafting program are unlimited. Once you get the hang of putting drawings on the computer, you are likely to enjoy it. You may even find computer drafting to be therapeutic.

Selecting Your Hardware

Selecting your hardware is not an easy task. While some people buy computer equipment to satisfy their curiosity about these electronic marvels or because they've got to have the latest technology, you should purchase what you need for use now and the foreseeable future. Computer technology changes so rapidly that we can only speculate on what the future holds. We do know one or two things —miniaturization and more power for less cost seem to characterize this industry. Big, bulky desktops gave way to laptops, which spawned smaller notebooks and pocket-size devices.

Getting an education in computers is rather simple these days. Go into a few electronics stores, fire up a computer or two, and ask the salesperson lots of questions. After several afternoons spent checking out these computers, learning the lingo, reading the specifications, and comparing prices, you probably will make the proper decision. The offerings between $1,000 and $2,000 are staggering, so you need to zero in on a manufacturer and decide on hard-drive capacity and monitor size and configuration (regular cathode tube or flat-panel LED screen).

Desktop computers have their advantages. The screens are larger, and if portability is of no concern, they will be less expensive than a comparable laptop. If you live in an area that is prone to power failures, remember that you may lose valuable information if you have not saved it when a power outage occurs. By contrast, battery-powered laptop computers will be unaffected by outages if they are operated in the battery as opposed to plug-in mode.

Laptop computers are very popular. These portable powerhouses are capable of the same tasks as many desktop units. However, the compact size and portability of laptops make them especially desirable for busy people on the move.

PRO POINTER

Although typing on the keyboard of a laptop can be tricky because the keys are so small and so closely placed, it is possible to attach a regular sized keyboard, making the laptop equal in most respects to the desk top, plus giving you the added feature of portability. With a spare battery or two you can have the best of both worlds.

CD Burners

Even moderately priced computers today include a CD burner program which allows you to create your own CDs. You can use this accessory to create marketing programs with pictures and text that you can distribute to prospective customers, download database information from your hard drive and transfer to the CD, or download music from the Internet (make sure it is legal!) and play it while you are doing some of your paperwork.

Wi-Fi

Wireless technology is the latest advance in electronic communication. Bluetooth is a new technology using radio waves to wirelessly connect digital devices to computers, cell phones, and other electronic equipment. Using a 2.4-gigahertz band of the unlicensed spectrum, Bluetooth (named for a Danish king in the 10th century) can send information much faster than a modem, but this connectivity works only on distances up to 30 feet. Wi-Fi is a wireless connection with the ability to connect devices as far as 300 feet apart and at speeds up to 11 megabits per second.

When Wi-Fi capability has been incorporated into handhelds, it will allow users to tap into their PCs and cell phones and gain access to the Internet and also to share documents with other team members. As more towers are built and ranges increase, voice and data communications will take another step toward being independent of cable connections.

Laser Printers vs. Ink Jet Printers

PRO POINTER

When considering the purchase of a laser printer as part of your ongoing printing expense, don't forget to include replacement printer cartridges—they can cost upwards of $65-$70.

Laser printers provide you with type clarity that matches any professional printing company's product. At one time these high-quality printers were in the $600-$1000 range, but like everything else in the field of electronics, prices have fallen dramatically. Color printers are more expensive than black-and-white laser printers, and while they certainly have a place in the office, a standard black-and-white printer is not only more affordable but also more practical. As your business expands and you need color printing for brochures and CAD work, you might consider upgrading.

Ink-jet printers are generally less costly than laser printers and offer high-quality printing, but don't forget to consider the replacement cost of the ink jets when you are shopping for one of these machines. Most of the major electronics manufacturers offer multi-purpose machines that act as printers, fax machines, and even scanners. Although the quality of each of the components may be slightly less than that of a machine specifically devoted to one task, the cost of these multi-purpose machines is such that they do deserve consideration when you are just starting to outfit your office.

CHAPTER NINE

Keeping Track of Your Cash

Money in, money out, accounts receivable, accounts payable—cash flow. Watching what comes in and what goes out is one of the most important—some builders say the most important—aspect of your business.

Keeping track of your cash may be more difficult than you think. This is especially true if you offer remodeling and repair services. While most new-home construction is tied to periodic payments from either the bank or the owner, repairs and short-term remodeling work require prompt billing and prompt payment. Suppliers and subcontractors have to be paid promptly, and unless you have a big bank account, you will depend upon payment from your customers in order to pay your suppliers and subcontractors. Extending credit to customers is risky business, but it is often a necessary evil when acting as a contractor.

PRO POINTER

If you don't learn to manage your money, the amount of money you make will have little bearing on your success. It is not a matter of how much money you make, but how much money you retain. Most contractors go out of business not because they don't know the construction business but they don't know business!

As a contractor, most of your accounts will be paid on time. Construction loans and draw disbursements make this possible. However, not all customers rely on construction loans. There are some people who pay with their own savings to have homes built and remodeled. There are several methods that you can employ to keep your business out of red ink.

Keeping Your Accounts Receivable and Payable Under Control

Payables can be the downfall of any business. If a business gets behind in its payables, the business is likely to spiral into a downward plunge. That supplier or subcontractor will begin sending letters that will become more demanding if ignored, and the threat of turning the account over for collection will follow. If the situation is not rectified, lawsuits and judgments will occur. In the end, the business will run out of money, and the only thing left will be a bad credit rating. If you think starting a new business is a challenge, you will be staggered by the difficulty in bringing a wounded business back.

Keep your accounts current. If you are unable to pay your bills, talk with your creditors. Credit managers are not ogres, but they do have a job to do. If you are honest and open with your suppliers, most of them will work with you. If you try to ignore the problem, it will only get worse. A healthy business needs a good credit history. If you have good credit, cherish it. If you don't, work to get a good rating.

Cash Flow

PRO POINTER

Having a business with a high net worth is not worth much if you don't have enough money to pay the bills. Cash flow is very important to a healthy business.

Cash flow is basically the amount of cash that comes in and the amount that goes out. If the amount coming in exceeds the amount owed, this is called positive cash flow. And, conversely, when the bills paid exceed the amount of money coming in, that is negative cash flow.

Positive cash flow is paramount to any business success. Paper profits are nice, but they don't pay the bills. Have you ever heard about the person who is land-rich and cash-poor? Well, it's true; I've had tremendous financial statements and nearly no cash.

I've seen a large number of businesses forced to the brink of bankruptcy, even though they had significant assets. These businesses held valuable assets, but couldn't convert them into cash quickly. There are liquid assets and non-liquid assets. Liquid assets are those that can be converted to cash quickly, say within 30 days. Although some non-liquid assets can be quickly converted to cash, they are generally sold at "fire-sale" prices. If a business becomes cash-poor, it is handicapped. A business without cash is like an army without ammunition. The cash may be on the way, but if the enemy attacks before it arrives, the business cannot defend itself. Regardless of your assets and business strength, if you don't have available cash, you are in trouble. As the old saying goes, "You've got to pay to play."

One of the biggest traps you can avoid is bad jobs. Doing work for a customer who is known to be a slow or poor payer should make you stop and think–do I need work that badly that I can risk not getting paid or getting paid 90 days later? I think a general rule of thumb is that if someone doesn't pay you in 30 days, they probably won't pay for at least 60 or 90 days. Jobs that result in slow or no pay can be your undoing. Don't get greedy. Greed is a major contributor to business failure. It is better to be patient, take a slow approach, and reach your goals gradually than to run flat out and not pay attention to your cash flow.

PRO POINTER

Some contractors will take a job at a very low price or no profit that often ends up with a loss, so rethink your decision in cases like that.

Looking Ahead to Financial Challenges

If you are looking ahead to financial challenges, you are on the right track. Even the best business will encounter setbacks from time to time, and if you are prepared for these slow periods, you have a better chance of survival. Business owners who learn to plan for the future are the ones who are around in the future. The years 2002-2004 were boom years for homebuilders in most parts of the country. The lowest interest rates in decades spawned a rush of refinancing, much of which was used to fuel the remodeling and renovation business. Low mortgage rates meant that many potential homeowners could afford to own a home, and present homeowners, using the equity in their homes and the hot market, were able to trade up and get a bigger house. But the housing industry is known for its cycles; boom or bust, and you need to anticipate cycles in your business plan. However, remodelers are rarely affected by the same swings that hurt builders. The economy historically cycles in upward and downward swings. If you can't cope with these changes, your business will not be around for the next change. Some unfortunate businesses get caught in the middle of a cycle. These businesses don't usually last to see the good times.

Sometimes I seem to have terrible luck with my timing. At one time, I had a building business that was building sixty homes a year. I enjoyed about eighteen months of a very prosperous business, but then the tax laws changed. What did tax laws have to do with my building business? About half the homes I was building were for partnerships. These homes were for tax shelters. I had devised a plan, with the help of tax attorneys and accountants, that was nearly perfect. The program was working without a flaw until the tax laws changed. When the tax advantages were removed from passive-income investments, my business suffered a great deal. This is only one example of how the economy has affected my businesses. I went through the recession of the early '80s, and I did okay in the economic downturn of the early '90s. I'm a survivor, and you can be, too.

PRO POINTER

With good planning and forecasting, you have a better-than-average chance of seeing light at the other end of the tunnel. By researching historical data, you can begin to see an economic pattern forming.

How can you overcome severe financial obstacles? The economy moves up and down; these cycles can be identified and charted. While there is no guarantee the cycles will repeat themselves, history indicates that they will. Interest rates are a key indicator—low rates spur construction activity, and, conversely, high interest rates put a damper on construction—so you need to keep in touch with what's happening in the financial industry. Check the business section of your local newspaper from time to time and tune in on those financial-news programs on cable TV. You must look ahead and plan. Having backup plans can make the difference between survival and failure.

Money Management

Without money management, you will not have a business for very long. Money management is critical to the successful operation of your business. If you are not experienced in money management, take some business courses at a local community college, read more business books, go to seminars, and do what it takes to get yourself educated in the intricate skill of managing money.

Contract Deposits

Getting deposit money from customers before the start of a job is not always easy, but it does ease the cash-flow burden. It is not unusual for remodeling or renovation contractors to receive cash deposits from their customers. A typical scenario finds the contractor getting one-third of the contract amount when the contract is signed, one-third at the halfway point of the job, and the balance upon completion. This, of course, is not the case with builders who are constructing new homes. Any deposit given on a new home is likely to be worth less than 10 percent of the contract amount. In fact, it is not unusual for homebuilders to begin work with a signed contract that does not include a deposit. But, many remodelers insist on advance deposits.

When advance deposits are given, such as in the case of remodeling jobs, contractors are working with money provided from customers. Without deposits, contractors must work with their own money and credit. This can put a serious strain on a contractor's bank account.

Many consumers have had bad experiences with contractors to whom they have given deposits. In the more extreme cases, the contractor never showed up. He skipped town with the customer's deposit. In other cases the contractor delayed the start of the work well beyond the date promised or failed to finish the work. Newspaper articles and talk shows often highlight these abuses. The public is very aware of the risks involved with giving money to someone before the work is done and is increasingly wary about giving a contractor deposit money. If you have established a good reputation, you are more likely to be able to obtain cash deposits from your customers. If the contractors don't get deposits, they are more exposed and must use their credit lines and cash to maintain their businesses. Because of the difficulty in obtaining deposits, make arrangements to work with your own money and a credit line. In addition, you will have to be more selective in your customers and make sure they will pay you. If you don't get deposits, you are at risk.

PRO POINTER

You should never take one person's deposit and use it on someone else's job. You also should never use deposit money for ordinary operating expenses.

Cash deposits enable you to do more business. However, it is important that you use the deposit money for the job the deposit was made on. All too often, contractors use deposits to make truck payments, rent payments, or even to buy materials for other jobs. This is referred to as "commingling of funds" and is a dangerous proposition. If you get into this habit, you will probably be out of business in less than a year.

Eliminating Subcontractor Deposits

Eliminating subcontractor deposits is another way to manage your money wisely. Just as homeowners are reluctant to give deposits to contractors, you should be selective in giving deposits to subcontractors. Once a sub has your deposit money, it can be difficult to get it back if he doesn't perform as expected. If a subcontractor asks you for a deposit, you can use the same reasons your customers give you for not wanting to give an upfront deposit.

When subcontractors are dependent upon your deposit to do jobs, they may be in financial trouble. If you give into these deposit requests, your money is at risk. Make it a rule to never give anyone money they haven't earned.

Stretch Your Money

If you learn to stretch your money, you can do more business. The more business you do, the more money you make. There are many ways to make your dollars go further. Let's see how you can increase the power of your cash:

- Rent expensive tools until you know you need them.
- Don't give subcontractors advance deposits.
- Collect job deposits from homeowners whenever possible.
- Make the best use of your time.
- Forecast financial budgets and stick to them.
- Buy in quantity for everyday items whenever feasible.
- Pay your supply bills early and take the discount.
- Put your operating capital in an interest-earning account.
- File extensions and pay your taxes late in the year.
- Don't overstock on inventory items.
- Keep employees working, not talking.
- Consider leasing big-ticket items.

These are only a few of the ways you can manage your money more effectively. If you study your business, you will find other ways to maximize you money.

Credit

While it is common for service companies to allow customers thirty days to pay their bills, this policy crunches your cash flow. Further, if you don't collect what's due you on the spot, you may have to wait 60 to 90 days for payment and in some cases never see your hard-earned money. Some people seem to neglect paying for services they receive. However, if you make an attempt to collect as soon as the job is done, most people will pay you. If you are on a service call, you should have a bill form handy, and when you are finished, fill in the labor rates and material costs, present them to the customer, and request a check. In fact, whoever takes the service call should advise the customer beforehand that you are requesting payment when the work is completed.

You may find that you must allow credit purchases to get business. But remember, working and supplying materials that you don't get paid for are worse than not working. If you allow credit purchases, be selective. After all, will banks loan you money without checking your credit history?

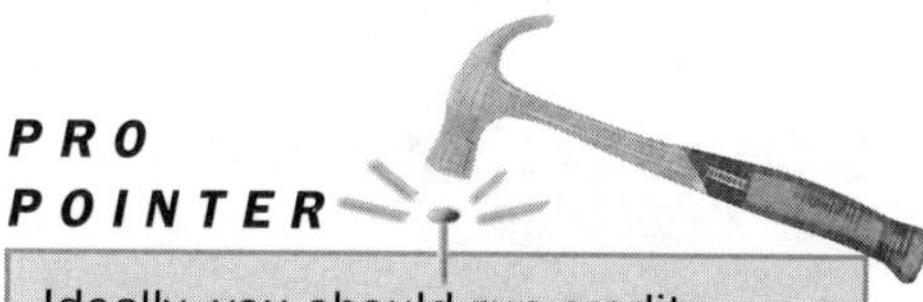

Ideally, you should run credit checks on all the customers who wish to establish a credit account.

Collecting Past-Due Accounts

Remember that if customers don't pay within 30 days, you may have a problem collecting your money. So the first rule of bill collecting is make sure you call all parties at the end of 30 days to find out when you can expect payment. That might jog the memory of some and might make others aware that you are on top of your collection work.

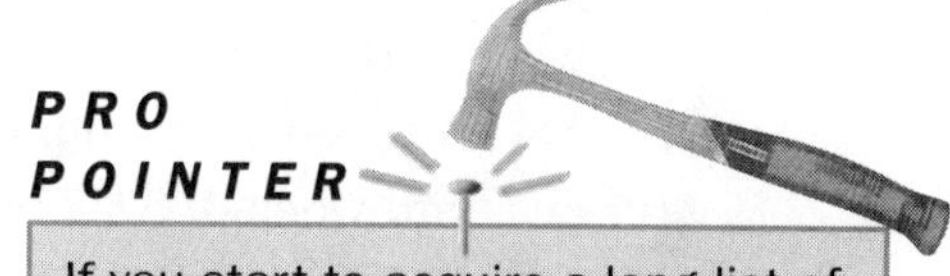

If you start to acquire a long list of accounts receivable, you are heading for deep trouble. Never let debts get too old.

Collecting past-due accounts is a job few people enjoy. As a business owner, you will no doubt have occasions to collect old money. This part of your job can get very frustrating. When you call to collect an old debt, you are likely to hear some very creative excuses. People who are behind in paying their bills will often lie to you. When you are listening to their excuses, don't get softhearted. Remember that if customers don't pay you, you cannot pay your bills.

As soon as a customer is in default, take action. Make a call first and request payment, follow up with a written notice indicating when payment was due, and request payment by a new date. If non-payment continues, you do have a couple of choices. There is Small Claims Court, where you can appeal to a judge by stating your case and requesting a judgment against the customer. There is a dollar limit on the disputed amount; in some states it is as high as $5,000. You must have all your documentation in hand: the type of work requested, any contracts, the bill that was presented to the customer, and your efforts to collect the bill (records of phone calls, copies of letters sent). Keep track of all phone calls to the customer. Be prepared for the customer to come up with reasons for non-payment—faulty workmanship, etc.—so you can be ready to counter any claims.

Construction Loans

PRO POINTER

You can always consult a local attorney for your options in collecting bad debts. But with attorney's fees in the $150 to $250 per hour range, it might not make much sense to pursue legal action for a $100 debt.

Constructions loans and home improvement loans vary quite a bit. I've worked with loans where I could submit invoices each week and receive checks. Some lenders have limited the disbursement process to once a month, which still isn't bad. Many construction loans pay out a percentage of the loan amount based on what work as been completed. Some loans are set up on an even pay-out basis, such as one-third when a house is under roof, one-third when drywall has been hung and taped, and one-third upon completion.

As a remodeler, you have to manage your money carefully. Part of doing this is knowing how and when you will be paid. I should clarify that we are not talking only about your profit payments. Many contractors pay their subcontractors out of their own funds and then get reimbursed by construction loans or home improvement loans obtained by customers. This is fine so long as everything is planned and goes as planned. But suppose you pay out $15,000 to subs and then find out that a draw disbursement is being held up due to some type of inspection?

In the industrial and commercial construction business it is common to have a "pay when paid" clause in subcontractor agreements. This clause, spelled out in legal language, states that the subcontractor will be paid within "X" number of days after the general contractor receives payment from the owner. Although this may be difficult to do as a remodeler who is just starting out, it characterizes one portion of the construction business; conserve your money by tying pay-outs to receipts whenever you can. It's easy to get into financial trouble as a contractor, so you need to keep close watch over what comes in and what goes out.

Keeping tabs on your money, such as petty cash, and the money that is due you (your receivables) can be a time-consuming process. Money is the oil that keeps a building machine going. Without cash, your business will crash. When you hang out your sign as a remodeler, you must have your financial ducks in a row.

PRO POINTER

If you're not good with budgets, schedules, and money, you need to give serious thought to hiring someone who is.

Taxes, accounting, legal considerations, and the paperwork that goes along with them are all important parts of your business. Each of these subjects is complex, and from time to time, you'll need to consult with a specialist in each field. After a while you'll begin to absorb some of the basics. Ask a lot of questions and you'll soon begin to feel more comfortable talking to your lawyer and accountant.

If you don't pay attention to your tax requirements, you will have long-term regrets. Severe penalties and interest charges add quickly to late tax payments, and sticking your head in the sand can cost you hundreds or thousands of dollars. The same can be true about the legal aspects of running your business. A poorly written contract can cause months of frustration, not to mention lost money.

PRO POINTER

Using and maintaining the proper paperwork can make all aspects of your business better. But when it comes to taxes and legal issues, organized paperwork is invaluable. If you ever have to go to court, you will learn the value of well-documented notes and agreements. An Internal Revenue Service (IRS) audit will prove the importance of keeping good records.

Professional help is not inexpensive, but it is better to pay an attorney to draft good legal documents than it is to pay to have improper documents defended in court. Your one-time cost to have an attorney prepare a contract and other legal forms will easily be repaid when you have to call on these documents to defend your position. The same can be said about accountants. A good certified public accountant (CPA) can save you money even after the professional fees you are charged.

Taxes

Taxes are a fact in business, and the tax code is complex. You need the services of an expert to help you interpret the law and figure out how to use its provisions to best advantage. For the average business owner, many tax advantages go unnoticed unless he or she is guided through this tax maze by an expert.

Unless you are a tax expert, you should consult someone who is. If you broke your leg, would you attempt to set the bone and apply a cast? Would you ever consider filling a cavity in your own tooth? Tax experts can save you money on the taxes you pay. When you talk with a CPA, you are talking to an expert. These professionals know the tax requirements like you know your business. Good record keeping is the key to controlling your taxes, but you need to consult with an accountant to get a clear understanding of the type of records you need to accumulate before your tax return is prepared.

PRO POINTER

If you prepare your own taxes, you may be giving the government much more money than you are obligated to pay.

If you wait until a month before tax time to meet with a tax specialist, there may not be much the expert can do for you, short of filing your return. However, if you consult with tax experts early in the year, you can manage your business to minimize the tax bite. Early consultations can result in significant savings for you and your business. A logical step is to meet with tax experts early and allow them to set a path for you to follow.

What type of planning can reduce your tax liabilities? Well, I'm not a tax expert, but I have found many ways, with the help of tax professionals, to reduce the money I spend in income taxes. If you take the time to meet with tax experts, you can find the best ways to run your business for maximum tax advantages.

PRO POINTER

Accountants do more than bookkeeping. A good CPA will be able to provide business planning that will result in lower tax consequences.

Interpretation of the tax code is an art, and CPAs are the artists. When you meet with these professionals, you might find numerous ways to save on your taxes. You might be told to lease vehicles instead of buying them. You may be shown how to keep a mileage log for your vehicle to maximize your tax deductions. A CPA might recommend a different type of structure for your business. For example, if you are operating as a standard corporation, the accountant might suggest that you switch to a subchapter S corporation to avoid double taxation.

If you work from home, your tax expert can show you how to deduct the area of your home that is used solely for business and deduct a portion of various household expenses. A tax specialist can show you how to defer your tax payments to a time when your tax rate may be lower. Investment strategies can be planned to make the most of your investment dollars. You can set up a viable retirement plan by talking with an authority on tax issues.

While tax experts will advise you on allowable expenses, they'll also tell you what expenses cannot be used as deductions. I see numerous cases where business owners are required to pay back taxes. Most of these business owners had no idea they owed taxes until they got a notice from the IRS to pay them. Deductions that they thought were deductible were not, and therefore their tax bill increased. With penalties and interest, this mistake can be a costly one. The best way to avoid unexpected tax bills is to make sure your taxes are filed and paid properly and promptly. All you have to do is to hire a professional to do the job for you.

Surviving an IRS Audit

If you are concerned about surviving an IRS audit, don't worry; I did it, and you can do it. For years, an audit was one of my largest fears. Even though I knew, or at least thought, that my tax filings were in order, I worried about the day I would be audited. It was a lot like dreading the semi-annual trip to the dentist. Then one day, it happened. I was notified that my tax records were going to be audited. My worst nightmare was becoming a reality.

Before the audit, I scrambled to gather old records and went over my tax returns for anything that might have been in error. I couldn't find any obvious problems. I went to my CPA and had him go over the information I would present in the audit. There were no blazing red flags to call attention to my tax return.

When I called the agent who was to perform my IRS audit, I was told that my return had been chosen at random. The person went on to say that there probably was nothing wrong with my return and that a percentage of returns are picked each year for audit. Gaining this information made me feel a little better, but not a lot.

On the day of the audit, my CPA represented me. I was not required to attend the meeting. Even though I wasn't at the meeting, my mental state was miserable that day. When my CPA called, he told me the meeting had gone well but that I had to provide further documentation for some of my deductions. These deductions were primarily travel expenses and books.

After digging through my records, I found most of the needed documentation. However, there were some receipts that I couldn't document. For example, I had written on receipts that they were for book purchases, but I hadn't listed the title of the book. The IRS wanted to know what books I had purchased. I couldn't remember what the titles were, so I put notes on the receipts to explain that I could not document the book titles. The travel expenses were easier to document; most of them had been paid for with credit cards. By finding the old receipts, I was able to remember where I had gone and why. After finding as much documentation as I could, the package was given back to my CPA.

Another meeting between the auditor and my accountant took place. I was expecting the worst and thought I would have to pay back-taxes on the items I couldn't identify properly. But to my surprise, I was given a clean bill of health. The auditor accepted my I-can't-remember receipts. Since most of my receipts were documented, my honesty prevailed in winning acceptance.

The end result was an experience I had never dreamed of. I always thought an audit would be horrible. However, I didn't have to attend the audit personally; my CPA did a great job, and I didn't have to pay any serious tax penalties. My audit was over without much pain. I can't say that all audits are this easy, but mine was.

Going through that audit convinced me of what I had believed for years: accurate records are the key to staying out of tax trouble. If my records had been misplaced or substantially incomplete, I could have been in a serious bind. However, my good business principles enabled me to survive the audit with relative ease.

PRO POINTER

Don't become upset because you have been notified of an audit. Turn your attention to keeping detailed records. Before taking a questionable deduction, check with a tax expert to confirm the viability of the deduction.

The Legal Side

The legal side of being in business can be perplexing. There are so many laws to abide by that the average person has trouble keeping up with them. There are laws that require certain posters to be displayed for employees. Contract law is an issue any contractor has to deal with. Laws pertaining to legal collection procedures for past-due accounts can affect your business. Discrimination laws come into play when you deal with employees. The list of laws that may affect your business could fill a small library. And most of them can affect how much money you make or keep.

Construction law is known as "case law". Current court rulings are based to a large extent on past court cases. Sometimes the law may be confusing, but as you meet with your lawyer and discuss matters you will become more familiar with some of the basic rules. For example, let's say you have installed a second bathroom in one of your jobs and that this was a change from the original contract. And let's say that you failed to get a signed change order from the owners indicating that they agreed to the change and the added cost. And, further, when the job is completed, they point to the original contract (which doesn't include the second bathroom nor the added cost) and tell you they are only paying the initial contract sum. Do you have any defense? Can you collect any money for the additional bathroom without that signed change order? Well, just talk to your attorney. There are two legal concepts, "unjust enrichment" and "quantum meruit," that may help you. Unjust enrichment means you can't get something for nothing. Your customers got a second bathroom and they must pay for it. The second term, "quantum meruit," is somewhat similar. It means, loosely, that the customer "benefited" by your actions (installing the bathroom), and there is an implied promise to pay.

This example is not meant to turn you into a guardhouse lawyer but merely to make you aware of the premise that if something seems unfair (your customer not wanting to pay) it probably is—but you need a lawyer to help you over these hurdles. And I'll bet you don't forget to get a signed change order the next time someone requests a change.

Business law is a vast subject, much too broad to cover in a single chapter. If you are not well versed in business law, spend some time studying the topic. You might find it necessary to go to seminars or college classes to gain the knowledge you need. Books that specialize in business law may be all the help you need. If you have specific questions, you can always confer with an attorney who concentrates on business law.

PRO POINTER

There are two legal concepts, "unjust enrichment" and "quantum meruit," that may help you.

Choosing Attorneys and Accountants

Choosing an attorney and an accountant is not always easy. Finding a professional with the experience, knowledge, and skill that you require can take some time. An individual attorney cannot possibly be versed in all aspects of the law. Accountants cannot be expected to know every

PRO POINTER

You must look for professionals who specialize in the type of service you require.

aspect of their profession, and they may not be acquainted with the provisions of the tax code.

Finding a specialist is easy. Most professionals list their specialties in the advertising they do. For example, an attorney who specializes in criminal law will often make this point clear in advertising. CPAs who concentrate on corporate accounts will direct their advertising to this form of service. However, finding professionals who advertise certain specialties is only the beginning. Ask some of your subcontractors or suppliers who have been in business for a number of years for the names of their attorneys.

Once you narrow the field to professionals working within the realms of your needs, you must further separate the crowd. Meet with prospective professionals and, in effect, interview them. You can begin your process of elimination with technical considerations. Make a list of what your known and expected needs are. Ask the selected professionals how they can help you with these needs. Inquire about the past performances and clients of the professionals. As you go down your list of questions, you will be able to begin weeding out the crowd.

Once you have covered all the technical questions, ask yourself some questions. How do you feel about the individual professional? Would you be comfortable going into a tax audit with this CPA? Are you willing to bet your business on the knowledge and courtroom prowess of this attorney? Are you comfortable talking with the individual? How you feel towards the professional as a person is important. You will very likely expose your deepest business secrets to your accountant and your attorney.

Another key factor to assess in choosing professionals is the ease with which you can understand them. The subject matter you will be discussing is complex and possibly foreign to you, and you need professionals who can decipher the cryptic information that confuses you and present it in an easy-to-understand manner.

PRO POINTER

If you are not comfortable with the professional you are working with, you will not get the most out of your business relationship.

Documenting Your Business Activity

Proper documentation of your business activity is an absolute must. Good documentation is essential to running a successful business. The documentation may be used to track sales, to keep up with changes in the market, to forecast the future, to defend yourself in court, or to substantiate your tax filing, just to name a few.

There are many good ways to document your business activity. A carbon-copy phone-message book is a simple yet effective way to log all your phone activity. Written contracts are instrumental in documenting your job obligations and responsibilities and the obligations and responsibilities of your customer. Change orders and addendums are the best way to document agreed-upon changes to the initial contract. And remember that changes generally affect both time and money. If your customers have requested a tight completion schedule because they may have sold their old house and are living with someone else until your house is finished, any major change can not only increase the cost but also extend the completion time. If the customer is not made aware of that fact, there could be a serious problem down the road. If your customer agrees that the added work requires additional time but wants you to meet the original deadline, you have an opportunity to estimate the amount of overtime you'll need to spend and tell them you will need to get reimbursed for this extra cost. Letters can be used to confirm phone conversations, create a paper record, and avoid confusion. Tape recorders can be an aid in remembering your daily duties. The reasons for documenting your business activity are many.

Inventory Logs

Inventory logs can be used to maintain current information on your inventory needs and document the materials and equipment you have purchased for your jobs. If you take materials out of your inventory, write in the log what the items were and where they were used. At the end of the week, go over your log and adjust your inventory figures. If you need to replace the inventory, you will know exactly what was used. If questions arise at tax time, you can identify where your inventory went. Something as simple as an inventory log can save you from lost money and time. At tax time you may need to document your starting inventory, additions, and deductions to arrive at an ending inventory for tax purposes, and the inventory log will help you in that regard.

PRO POINTER

Receipts will serve as documentation for your deductions. If the receipt does not state clearly what the purchase was, write in the details. You should include what was purchased and what it was for. If the item was for a particular job, write the job name on the receipt. In a tax audit, your documented receipts may mean the difference between an easy audit and having to pay back taxes.

Contracts, Change Orders, and Related Paperwork

Contracts, change orders, and related essential paperwork will help keep your business on the right track. While most people abhor doing paperwork, it is recognized as a necessary part of doing business. The use of computers has helped to reduce the amount of paper used and made the task easier, but even computers cannot eliminate the need for accurate paperwork. To understand the need for so much clerical work, let's take a closer look at the various needs of business owners.

Contracts

Contracts are a way of life for contractors. The contracts that a contractor enters into are the lifeblood of the business. There are two types of contracts–oral and written. Oral contracts are legal, but they are usually unenforceable (although under certain circumstances the courts have upheld some verbal contracts). Written contracts, on the other hand, are the best form of protection against misinterpretation of the terms and conditions of the transaction between you and your customer or subcontractor. .

A written contract should be used for every job except small repairs where a more condensed version can be prepared. Contracts are there to protect you because they will provide physical evidence of the understanding between you and the customer. There are certain essential elements in every good contract:

- The scope of the work being contracted, whether it is a set of plans and specifications, a list of work items to be undertaken, or a written description of the work that you and your customer have agreed upon
- The cost of the work that has been accepted for the work described and the method of payment and dates when payments are to be made; references for deposits would be included in this portion of the contract
- The time frame in which the work will take place–the start date, duration, and completion date
- Other terms relating to changes in the work, acceptable delays, and miscellaneous provisions unique to this project and how they will be handled
- Signatures of both buyer and seller attesting to acceptance of the agreement, along with the date the contract was signed

Professional help in preparing a basic contract format(s) that you will use throughout your business is money well spent. You will need one contract format for subcontractors and one for homebuyers.

Your Company Name
Your Company Address
Your Company Phone Number

REMODELING CONTRACT

This agreement, made this ________ day of __________, 20___, shall set forth the whole agreement, in its entirety, between Contractor and Customer.

Contractor: Your Company Name, referred to herein as Contractor.

Customer: ______________________________________, referred to herein as Customer.

Job name: ____

Job location: ____

The Customer and Contractor agree to the following:

SCOPE OF WORK

Contractor shall perform all work as described below and provide all material to complete the work described below. All work is to be completed by Contractor in accordance with the attached plans and specifications. All material is to be supplied by Contractor in accordance with the attached plans and specifications. Said attached plans and specifications have been acknowledged and signed by Contractor and Customer.

A brief outline of the work is as follows, and all work referenced in the attached plans and specifications will be completed to the Customer's reasonable satisfaction. The following is only a basic outline of the overall work to be performed: ____________________________________

____

____

____

____

____

____

____

____

____

(Page 1 of 3. Please initial _____.)

FIGURE 9.1 Example of a remodeling contract. *(continued on next page)*

COMMENCEMENT AND COMPLETION SCHEDULE

The work described above shall be started within _______ (_____) days of verbal notice from Customer; the projected start date is ___________. The Contractor shall complete the above work in a professional and expedient manner, by no later than ______ (____) days from the start date. Time is of the essence regarding this contract. No extension of time will be valid without the Customer's written consent. If Contractor does not complete the work in the time allowed, and if the lack of completion is not caused by the Customer, the Contractor will be charged ________ ($____) dollars per day, for every day work is not finished beyond the completion date. This charge will be deducted from any payments due to the Contractor for work performed.

CONTRACT SUM

The Customer shall pay the Contractor for the performance of completed work, subject to additions and deductions, as authorized by this agreement or attached addendum. The contract sum is ______________________________________, ($______________).

PROGRESS PAYMENTS

The Customer shall pay the Contractor installments as detailed below, once an acceptable insurance certificate has been filed by the Contractor, with the Customer.

Customer will pay Contractor a deposit of ______________________________ ________________________, ($___________), when work is started.

Customer will pay ____, ($____________), when all rough-in work is complete.

Customer will pay ____, ($______________) when work is __________ (_____%) percent complete.

Customer will pay ____, ($____________) when all work is complete and accepted.

All payments are subject to a site inspection and approval of work by the Customer. Before final payment, the Contractor, if required, shall submit satisfactory evidence to the Customer, that all expenses related to this work have been paid and no lien risk exists on the subject property.

WORKING CONDITIONS

Working hours will be ______ a.m. through ______ p.m., Monday through Friday. Contractor is required to clean work debris from the job site on a daily basis and to leave the site in a clean and neat condition. Contractor shall be responsible for removal and disposal of all debris related to the job description.

CONTRACT ASSIGNMENT

Contractor shall not assign this contract or subcontract the whole of this contract without the written consent of the Customer.

(Page 2 of 3. Please initial _____.)

FIGURE 9.1 *(continued)* Example of a remodeling contract.

(continued on next page)

LAWS, PERMITS, FEES, AND NOTICES

Contractor is responsible for all required laws, permits, fees, or notices required to perform the work stated herein.

WORK OF OTHERS

Contractor shall be responsible for any damage caused to existing conditions. This shall include work performed on the project by other contractors. If the Contractor damages existing conditions or work performed by other contractors, said Contractor shall be responsible for the repair of said damages. These repairs may be made by the Contractor responsible for the damages or another contractor, at the sole discretion of Customer.

The damaging Contractor shall have the opportunity to quote a price for the repairs. The Customer is under no obligation to engage the damaging Contractor to make the repairs. If a different contractor repairs the damage, the Contractor causing the damage may be back charged for the cost of the repairs. These charges may be deducted from any monies owed to the damaging Contractor.

If no money is owed to the damaging Contractor, said Contractor shall pay the invoiced amount within _______ (_____) business days. If prompt payment is not made, the Customer may exercise all legal means to collect the requested monies. The damaging Contractor shall have no rights to lien the Customer's property for money retained to cover the repair of damages caused by the Contractor. The Customer may have the repairs made to his or her satisfaction.

WARRANTY

Contractor warrants to the Customer all work and materials for one year from the final day of work performed.

INDEMNIFICATION

To the fullest extent allowed by law, the Customer shall indemnify and hold harmless the Contractor and all of his or her agents and employees from and against all claims, damages, losses, and expenses.

This Agreement entered into on __________________, 20____ shall constitute the whole agreement between Customer and Contractor.

_______________________________ _______________________________

Customer Date Contractor Date

____

Customer Date

(Page 3 of 3)

FIGURE 9.1 ***(continued)*** Example of a remodeling contract.

SAMPLE REMODELING CONTRACT

RICHARD & RHONDA SMART
180 HOMEOWNER LANE
WIZETOWN, OH 99897
(102) 555-6789

REMODELING CONTRACT

This agreement, made this 12th day of May, 2004, shall set forth the whole agreement, in its entirety, between Contractor and Homeowner
Contractor: Generic General Contractors, referred to herein as Contractor.
Owner: Richard and Rhonda Smart, referred to herein as Homeowner.

Job name: Smart Kitchen Remodel

Job location: 180 Homeowner Lane, Wizetown, OH

The Homeowner and Contractor agree to the following:

SCOPE OF WORK

Contractor shall perform all work as described below and provide all material to complete the work described below:

All work is to be completed by Contractor in accordance with the attached plans and specifications. All material is to be supplied by Contractor in accordance with attached plans and specifications. Said attached plans and specifications have been acknowledged and signed by Homeowner and Contractor.

A brief outline of the work is as follows, this work is only part of the work, and all work referenced in the attached plans and specifications will be completed to the Homeowner's satisfaction. The following is only a basic outline of the overall work to be performed:

REMOVE EXISTING KITCHEN CABINETS
REMOVE EXISTING KITCHEN FLOOR COVERING AND UNDERLAYMENT
REMOVE EXISTING KITCHEN SINK AND FAUCET
REMOVE EXISTING COUNTERTOP
REMOVE EXISTING ELECTRICAL FIXTURES, SWITCHES, AND OUTLETS
REMOVE EXISTING KITCHEN WINDOW
SUPPLY AND INSTALL NEW KITCHEN CABINETS
SUPPLY AND INSTALL NEW KITCHEN SINK AND FAUCET
SUPPLY AND INSTALL NEW KITCHEN UNDERLAYMENT AND FLOOR COVERING
SUPPLY AND INSTALL NEW KITCHEN COUNTERTOP

(page 1 of 4 initials ______)

FIGURE 9.2 Example of a remodeling contract. *(continued on next page)*

SAMPLE REMODELING CONTRACT (continued)

SUPPLY AND INSTALL NEW ELECTRICAL FIXTURES, SWITCHES, AND OUTLETS
SUPPLY AND INSTALL NEW KITCHEN GREENHOUSE WINDOW
PATCH, SAND, PRIME AND PAINT WALLS, CEILING, AND TRIM
COMPLETE ALL WORK IN STRICT COMPLIANCE WITH ATTACHED PLANS AND SPECIFICATIONS, ACKNOWLEDGED BY ALL PARTIES.

COMMENCEMENT AND COMPLETION SCHEDULE

The work described above shall be started within three days of verbal notice from Homeowner, the projected start date is 6/20/04. The Contractor shall complete the above work, in a professional and expedient manner, by no later than twenty days from the start date. Time is of the essence regarding this contract. No extension of time will be valid without the Homeowner's written consent. If Contractor does not complete the work in the time allowed, and if the lack of completion is not caused by the Homeowner, the Contractor will be charged One Hundred Dollars, ($100.00), per day, for every day work is not finished, beyond the completion date. This charge will be deducted from any payments due to the Contractor for work performed.

CONTRACT SUM

The Homeowner shall pay the Contractor for the performance of completed work, subject to additions and deductions, as authorized by this agreement or attached addendum. The Contract Sum is Ten Thousand Three Hundred Dollars, ($10,300.00).

PROGRESS PAYMENTS

The Homeowner shall pay the Contractor installments as detailed below, once an acceptable insurance certificate has been filled by the Contractor, with the Homeowner:

Homeowner will pay Contractor a deposit of One Thousand Five Hundred Dollars ($1,500.00), when demolition work is started.

Homeowner will then pay Two Thousand Dollars, ($2,000.00), when all demolition and rough-in work is complete.

Homeowner will pay Three Thousand Dollars, ($3,000.00), when walls have been painted, and cabinets, countertops, and flooring have been installed.

(page 2 of 4 initials_______)

FIGURE 9.2 *(continued)* Example of a remodeling contract.

(continued on next page)

SAMPLE REMODELING CONTRACT (continued)

Homeowner will pay Three Thousand, Two Hundred Dollars ($3,200.00) when all work is complete. Homeowner will pay the final Five Hundred Fifty Dollars ($550.00) within thirty days of completion, if no problems occur and are left uncorrected.

All payments are subject to a site inspection and approval of work by the Homeowner. Before final payment, the Contractor, if required, shall submit satisfactory evidence to the Homeowner, that all expenses, related to this work have been paid and no lien risk exists on the subject property.

WORKING CONDITIONS

Working hours will be 8:00 a.m. through 4:30 a.m., Monday through Friday. Contractor is required to clean their work debris from the job site on a daily basis, and leave the site in a clean and neat condition. Contractor shall be responsible for removal and disposal of all debris related to their job description.

CONTRACT ASSIGNMENT

Contractor shall not assign this contract or further subcontract the whole of this subcontract without the written consent of the Homeowner.

LAWS, PERMITS, FEES, AND NOTICES

Contractor is responsible for all required laws, permits, fees, or notices, required to perform the work stated herein.

WORK OF OTHERS

Contractor shall be responsible for any damage caused to existing conditions. This shall include work performed on the project by other contractors. If the Contractor damages existing conditions or work performed by other contractors, said Contractor shall be responsible for the repair of said damages. These repairs may be made by the Contractor responsible for the damages or another contractor, at the sole discretion of Homeowner.

The damaging Contractor shall have the opportunity to quote a price for the repairs. The Homeowner is under no obligation to engage the damaging Contractor to make the repairs.

(page 3 of 4 initials ______)

FIGURE 9.2 *(continued)* Example of a remodeling contract.

(continued on next page)

SAMPLE REMODELING CONTRACT (continued)

If a different contractor repairs the damage, the Contractor causing the damage may be back charged for the cost of the repairs. These charges may be deducted from any monies owed to the damaging Contractor by the Homeowner.

If no money is owed to the damaging Contractor, said Contractor shall pay the invoiced amount, from the Homeowner, within seven business days. If prompt payment is not made, the Homeowner may exercise all legal means to collect the requested monies.

The damaging Contractor shall have no rights to lien the Homeowner's property for money retained to cover the repair of damages caused by the Contractor. The Homeowner may have the repairs made to their satisfaction.

WARRANTY

Contractor warrants to the Homeowner all work and materials for one year from the final day of work performed.

INDEMNIFICATION

To the fullest extent allowed by law, the Contractor shall indemnify and hold harmless the Homeowner and all of their agents and employees from and against all claims, damages, losses and expenses.

This Agreement entered into on May 12, 2004 shall constitute the whole agreement between Homeowner and Contractor.

Homeowner ________________ Contractor ________________

(page 4 of 4 initials______)

FIGURE 9.2 ***(continued)*** Example of a remodeling contract.

SAMPLE OWNER-SUPPLIED SUBCONTRACT AGREEMENT

RICHARD & RHONDA SMART
180 HOMEOWNER LANE
WIZETOWN, OH 99897
(102) 555-6789

SUBCONTRACT AGREEMENT

This agreement, made this 25th day of July, 2004, shall set forth the whole agreement, in its entirety, between Contractor and Subcontractor.

Contractor: Richard & Rhonda Smart, referred to herein as Contractor.
Job location: 180 Homeowner Lane, Wizetown, OH
Subcontractor: Wild Bill's Painting company, referred to herein as Subcontractor.

The Contractor and Subcontractor agree to the following:

SCOPE OF WORK

Subcontractor shall perform all work as described below and provide all material to complete the work described below:

Subcontractor shall supply all labor and material to complete the work according to the attached plans and specifications. These attached plans and specifications have been initialed and signed by all parties. The work shall include, but is not limited to, the following:

(1) Scrape all painted surfaces in the family room, living room, and bedrooms.
(2) Fill all cracks and holes with joint compound.
(3) Sand painted surfaces as needed and prepare all painted surfaces for new paint.
(4) Provide protection from paint or other substance spillage.
(5) Move and replace any obstacles, furniture, or other items in the area to be painted.
(6) Prime all surfaces to be painted with an approved primer.
(7) Paint all existing painted surfaces with two coats of a Latex paint, color number LT1689.
(8) Remove any excess paint from window glass or other areas not intended to be painted.
(9) Complete all work in strict compliance with the attached plans and specifications.

(Page 1 of 3 initials ___)

FIGURE 9.3 Example of an owner-supplied subcontract agreement.

(continued on next page)

SAMPLE OWNER-SUPPLIED SUBCONTRACT (continued)

COMMENCEMENT AND COMPLETION SCHEDULE

The work described above shall be started within three days of verbal notice from Contractor, the projected start date is 8/20/04. The Subcontractor shall complete the above work in a professional and expedient manner by no later than twenty days from the start date. Time is of the essence in this Subcontract. No extension of time will be valid without the Contractor's written consent. If Subcontractor does not complete the work in the time allowed, and if the lack of completion is not caused by the Contractor, the Subcontractor will be charged Fifty Dollars ($50.00) per day, for every day work extends beyond the completion date. This charge will be deducted from any payments due to the Subcontractor for work performed.

CONTRACT SUM

The Contractor shall pay the Subcontractor for the performance of completed work, subject to additions and deductions, as authorized by this agreement or attached addendum. The Contract Sum is Two Thousand Dollars ($2,000.00).

PROGRESS PAYMENTS

The Contractor shall pay the Subcontractor installments as detailed below, once an acceptable insurance certificate has been filled by the Subcontractor with the Contractor:

Contractor shall pay the Subcontractor Five Hundred Dollars ($500.00), when materials are delivered and preparation work is started.

Contractor shall pay the Subcontractor Five Hundred Dollars ($500.00), when preparation work is complete and painting is started.

Contractor shall pay the Subcontractor Eight Hundred Dollars ($800.00), when all work is complete and approved by the Contractor.

Contractor shall pay the Subcontractor Two Hundred Dollars ($200.00), thirty days after completion and acceptance of work, if no deficiencies are found in materials or workmanship during the thirty day period.

All payments are subject to a site inspection and approval of work by the Contractor. Before final payment, the Subcontractor, shall submit satisfactory evidence to the Contractor that no lien risk exists on the subject property.

(Page 2 of 3 initials ___)

FIGURE 9.3 *(continued)* Example of an owner-supplied subcontract agreement.

(continued on next page)

SAMPLE OWNER-SUPPLIED SUBCONTRACT (continued)

WORKING CONDITIONS

Working hours will be 8:00 a.m. through 4:30 a.m., Monday through Friday. Subcontractor is required to clean his work debris from the job site on a daily basis and leave the site in a clean and neat condition. Subcontractor shall be responsible for removal & disposal of all debris related to his job description.

CONTRACT ASSIGNMENT

Subcontractor shall not assign this contract or further subcontract the whole of this subcontract, without the written consent of the Contractor.

LAWS, PERMITS, FEES, AND NOTICES

Subcontractor shall be responsible for all required laws, permits, fees, or notices, required to perform the work stated herein.

WORK OF OTHERS

Subcontractor shall be responsible for any damage caused to existing conditions or other contractor's work. This damage will be repaired, and the Subcontractor charged for the expense and supervision of this work. The Subcontractor shall have the opportunity to quote a price for said repairs, but the Contractor is under no obligation to engage the Subcontractor to make said repairs. If a different subcontractor repairs the damage, the Subcontractor may be back-charged for the cost of the repairs.

Any repair costs will be deducted from any payments due to the Subcontractor, if any exist. If no payments are due the Subcontractor, the Subcontractor shall pay the invoiced amount within ten days.

WARRANTY

Subcontractor warrants to the Contractor, all work and materials for one year from the final day of work performed.

INDEMNIFICATION

To the fullest extent allowed by law, the Subcontractor shall indemnify and hold harmless the Owner, the Contractor, and all of their agents and employees from and against all claims, damages, losses and expenses. This agreement, entered into on July 25, 2004, shall constitute the whole agreement between Contractor and Subcontractor.

______________________	______________________
Contractor	Subcontractor
Richard B. Smart	Rhonda M. Smart

(Page 3 of 3 initials ___)

FIGURE 9.3 Example of an owner-supplied subcontract agreement.

SAMPLE SUBCONTRACTOR-SUPPLIED CONTRACT

Anytime Plumbing & Heating
126 OCEAN STREET
BEACHTOWN, ME 00390
(000) 123-4567

PROPOSAL CONTRACT

TO: Mr. and Mrs. Homeowner Date: 8/17/04
ADDRESS: 52 Your Street Beachtown, ME 0039 PHONE: (000) 123-9876
JOB LOCATION: Same JOB PHONE: Same PLANS: Drawn by ACS, 4/14/04

ANYTIME PLUMBING & HEATING PROPOSES THE FOLLOWING:

Anytime Plumbing & Heating will supply and or coordinate all labor and material for the work referenced below:

PLUMBING

Supply and install a 3/4", type "L", copper water main from ten feet outside the foundation, to the location shown on the attached plans for the new addition.

Supply and install a 4", schedule 40, sewer main to the addition, from ten feet outside the foundation, to the location shown on the plans.

Supply and install schedule 40, steel gas pipe from the meter location, shown on the plans, to the furnace, in the attic, as shown on the plans.

Supply and install the following fixtures, as per plans, except as noted:

1 ABC Venus one piece, fiberglass, tub/shower unit, in white.
1 CF 007_222218 chrome tub/shower faucet.
1 ABC 900928 water closet combination, in white.
1 CBA 111 cultured marble, 30" vanity top, in white.
1 CF 005-95011 chrome lavatory faucet.
1 PKT 11122012 stainless steel, double bowl, kitchen sink.
1 CF 908001 chrome kitchen faucet.
1 DFG 62789 52 gallon, electric, 5 year warranty, water heater.
1 WTFC 20384 frost proof, anti-siphon silcock.
1 AWD 90576 3/4" backflow preventer.
1 FT66754W white, round front, water closet seat.
1 plastic washer box, with hose bibs.
Connect owner-supplied dishwasher.

All fixtures are subject to substitution with fixtures of similar quality, at Anytime Plumbing & Heating's discretion.

All water distribution pipe, after the water meter, will be Pex tubing, run under the slab. This is a change from the specifications and plans, in an attempt to reduce cost.

(Page 1 of 3) *Initials*__________

FIGURE 9.4 Example of a subcontractor-supplied contract. *(continued on next page)*

SAMPLE SUBCONTRACTOR-SUPPLIED CONTRACT (continued)

If water pipe is run as specified in the plans, the pipe will be, type "L" copper and there will be additional cost. Any additional cost will be added to the price listed in this proposal.

All waste and vent pipes will be schedule 40 PVC.

Anytime Plumbing & Heating will provide for trenching the inside of the foundation, for underground plumbing. If the trenching is complicated by rock, unusual depth, or other unknown factors, there will be additional charges. These charges will be for the extra work involved in the trenching.

All plumbing will be installed to comply with state and local codes. Plumbing installation may vary from the plumbing diagrams drawn on the plans.

Anytime Plumbing & Heating will provide roof flashings for all pipes penetrating the roof, but will not be responsible for their installation.

All required holes in the foundation will be provided by others.

All trenching, outside of the foundation, will be provided by others.

All gas piping, outside the structure, will be provided by others.

The price for this plumbing work will be, Four Thousand, Eighty Seven Dollars ($4,087.00).

HEATING

Anytime Plumbing & Heating will supply and install all duct work and registers, as per plans.

Anytime Plumbing & Heating will supply and install a BTDY-P5HSD12NO7501 gas fired, forced hot air furnace. The installation will be, as per plans. The homeowner will provide adequate access for this installation.

Venting for the clothes dryer and exhaust fan is not included in this price. The venting will be done at additional charge, if requested.

No air conditioning work is included.

The price for the heating work will be Three Thousand, Eight Hundred Dollars ($3,800.00).

Any alterations to this contract will only be valid, if in writing and signed by all parties. Verbal arrangements will not be binding.

PAYMENT WILL BE AS FOLLOWS:

Contract Price of: Seven Thousand, Eight Hundred Eighty-Seven Dollars ($7,887.00), to be paid; one third ($2,629.00) at the signing of the contract. One third ($2,627.00) when the plumbing and heating is roughed-in. One third ($2,629.00) when work is completed. All payments shall be made within five business days of the invoice date.

*(Page 2 of 3) Initials*___________

FIGURE 9.4 *(continued)* Example of a subcontractor-supplied contract.

(continued on next page)

SAMPLE SUBCONTRACTOR-SUPPLIED CONTRACT (continued)

If payment is not made according to the terms above, Anytime Plumbing & Heating will have the following rights and remedies. Anytime Plumbing & Heating may charge a monthly service charge of one percent (1%), twelve percent (12%) per year, from the first day default is made. Anytime Plumbing & Heating may lien the property where the work has been done. Anytime Plumbing & Heating may use all legal methods in the collection of monies owed to Anytime Plumbing & Heating. Anytime Plumbing & Heating may seek compensation, at the rate of $50.00 per hour, for their employees attempting to collect unpaid monies. Anytime Plumbing & Heating may seek payment for legal fees and other costs of collection, to the full extent that law allows.

If Anytime Plumbing & Heating is requested to send men or material to a job by their customer or their customer's representative, the following policy shall apply. If a job is not ready for the service or material requested, and the delay is not due to Anytime Plumbing & Heating's actions, Anytime Plumbing & Heating may charge the customer for their efforts in complying with the customer's request. This charge will be at a rate of $50.00 per hour, per man, including travel time.

If you have any questions or don't understand this proposal, seek professional advice. Upon acceptance this becomes a binding contract between both parties.

Respectfully submitted,

H. P. Contractor
Owner

PROPOSAL EXPIRES IN 30 DAYS, IF NOT ACCEPTED BY ALL PARTIES

ACCEPTANCE
We the undersigned, do hereby agree to and accept all the terms and conditions of this proposal. We fully understand the terms and conditions and hereby consent to enter into this contract.

Anytime Plumbing & Heating	Customer #1
by________________________	________________________
Title______________________	Date______________________
Date______________________	Customer #2
	________________________
	Date______________________

(Page 3 of 3)

FIGURE 9.4 *(continued)* Example of a subcontractor-supplied contract.

Addendums

PRO POINTER

The blank contract forms sold at office-supply stores are not the way to go—they don't have all of the protection built in that you will need.

Addendums are extensions of a contract. Particularly when your contract with the owner includes a specific set of plans and changes to the plans are made, an addendum to the contract will be prepared to include those changes. With the issuance of an addendum, the contract sum will generally change, either upward–if more work has been added–or downward–if some work has been deleted. The addendum to the contract formalizes these changes and usually triggers a change order.

Change Orders

Change orders are written agreements that are used when a change is made to a previous contract. For example, if you were contracted to paint a house white and the customer asked you to change the color to beige, you should use a change order. If you paint the house beige without a change order, you could be found in breach of your contract, since the contract called for the house to be painted white. It may seem unlikely that the customer would sue you for doing what you were told, but it is possible. Without a written change order, you would be at the mercy of the court. If the court accepted the contract as proof that the house was to be painted white, and it probably would, you would be in trouble.

When changes to the scope of work are requested by your customer, they need to be documented by a change order. A full description of the change needs to be spelled out. Change orders can impact both time and cost or, in the case stated above regarding paint, have little or no impact on time or cost. But in either case they need to be documented by the issuance of a change order. When the owner requests, let's say, a change from resilient flooring to quarry tile in the kitchen, it is important to prepare a change order specifying this change, the cost, and any impact on construction time, and get it signed before you proceed. Too often a customer will tell you to institute a change and you don't inform what the cost will be, but when you are finished and present the bill he is shocked. His reply will be ,"If I knew it was going to be that much, I never would have told you to go ahead–I just think that price is outrageous." You'll then have to compromise (translated–reduce) on the price and end up with a disgruntled customer.

CONTRACTOR EXCLUSION ADDENDUM

This exclusion addendum shall become an integral part of the contract, dated August 15, 2004, between the customer, Mr. & Mrs. J. P. Homeowner, and the contractor, Anytime Plumbing & Heating, for the work to be performed on Mr. & Mrs. J. P. Homeowner's residence. The residence and job location is located at 135 Hometown street, in the city of Wahoo, State of Vermont. The following exclusions are the only exclusions to be made. These exclusions shall become a part of the original contract and may not be altered again without written authorization from all parties.

THE EXCLUSIONS

Anytime Plumbing & Heating Company will provide for trenching the inside of the foundation for underground plumbing. If the trenching is complicated by rock, unusual depth, or other unknown factors, there will be cause for additional charges. The homeowner shall have the right to contract a different company to remove these obstacles, without affecting the agreed upon price between the homeowner and Anytime Plumbing & Heating Company. Anytime Plumbing & Heating Company will provide these services, if requested, in writing, at an hourly rate of $50.00 per hour.

Anytime Plumbing & Heating Company will provide roof flashings for all pipes penetrating the roof, but will not be responsible for their installation.

All trenching outside of the foundation will be provided by others.

The above exclusions are the only exclusions allowed to the contract between the two parties. There are no verbal agreements. This two-page document represents the whole agreement between the parties, as it pertains to exclusions.

ACCEPTANCE

We the undersigned do hereby agree to and accept all the terms and conditions of this addendum. We fully understand the terms and conditions and hereby consent to enter into this contract addendum.

Anytime Plumbing & Heating Company Customer #1

by ____________________________

Title ____________________________ Date ________

by ____________________________ Customer #2

Date _______

FIGURE 9.5 Example of a contractor exclusion addendum.

BID ADDENDUM
REQUEST FOR SUBSTITUTIONS

CUSTOMER NAME: Mr. & Mrs. J. P. Homeowner
CUSTOMER ADDRESS: 192 Hometown Street
CUSTOMER CITY/STATE/ZIP: Yooho City, NA 93001
CUSTOMER PHONE NUMBER: (000) 756-3333
JOB LOCATION: Same
PLANS & SPECIFICATIONS DATED: June 10, 2004
BID REQUESTED FROM: Mid Range Suppliers
SUPPLIER ADDRESS: 42 Supplier Street
CONTACT PERSON: Liz Materialwoman, Manager
DATE: July 25, 2004
TYPE OF WORK: Remodeling

THE FOLLOWING ITEMS ARE BEING SUBSTITUTED FOR THE ITEMS SPECIFIED IN THE ATTACHED PLANS AND SPECIFICATIONS:

Roof shingles-The brand specified is not readily available. Our proposed substitute is product number 2246 form WXYZ company. The type, color, and general characteristics are very similar.

Siding-The brand requested is not available through our distribution network. It can be special ordered, but this requires payment prior to order placement. A proposed substitute is product number 4456 from ABEC company. The color and general features are essentially the same as the requested siding.

________________________ ________________________
Contractor Date Customer Date

FIGURE 9.6 Example of a bid addendum request for substitutions.

Your Company Name
Your Company Address
Your Company Phone and Fax Numbers

REQUEST FOR SUBSTITUTIONS

Customer name: ____________________

Customer address: ____________________

Customer city/state/zip: ____________________

Customer phone number: ____________________

Job location: ____________________

Plans and specifications dated: ____________________

Bid requested from: ____________________

Type of work: ____________________

The following items are being substituted for the items specified in the attached plans and specifications: ____________________

____

____

____

____

____

Please indicate your acceptance of these substitutions by signing below.

____

Contractor Date Customer
Date

____

FIGURE 9.7 Example of a request for substitutions form.

Your Company Name
Your Company Address
Your Company Phone and Fax Number

CHANGE ORDER

This change order is an integral part of the contract dated ___________, between the customer _________________________, and the contractor, ____, for the work to be performed. The job location is _____________. The following changes are the only changes to be made. These changes shall now become a part of the original contract and may not be altered again without written authorization from all parties.

Changes to be as follows:

____

____

____

____

____

These changes will increase / decrease the original contract amount. Payment for theses changes will be made as follows: ____. The amount of change in the contract price will be ____________________________________ ($___________). The new total contract price shall be ________________________________ ($___________).

The undersigned parties hereby agree that these are the only changes to be made to the original contract. No verbal agreements will be valid. No further alterations will be allowed without additional written authorization, signed by all parties. This change order constitutes the entire agreement between the parties to alter the original contract.

______________________________ ______________________________
Customer Contractor

______________________________ ______________________________
Date Date

____
Customer

____
Date

FIGURE 9.8 Example of a change order form.

Service Orders

PRO POINTER

When a change is requested, get it in writing—with the price—and have the customer sign it.

Service orders have several functions; they are usually used to document the acceptance of some minor work repair, but these tickets can also serve as a mini-invoice—with the addition of some additional items. A service order can document the labor hours spent on the job and include the materials or parts required. Merely adding the hourly rate and multiplying by the number of hours, you have the total labor costs. When you add in the materials and your overhead and profit, you can present the customer with an invoice and ask for a check before you leave the job. You can also use these service tickets to stock your parts shelf. When you see that you are using lots of one material or part, you can create a meaningful inventory of that material or part. In small print you can include guarantees, liability issues, restrictions, and so forth. While you are at your lawyer's office discussing a contract format, ask her to review a sample service order.

As a by-product, when your employees turn these tickets into the office, you can get a good idea of their productivity and also use this information for pricing out similar jobs in the future.

If you are going to provide routine maintenance and repair services, you will need to use a service order. Service orders are the small tickets that customers are asked to sign, generally after the work is done, to acknowledge that the work is satisfactory. Most customers are not asked to sign the service order until the work is complete, but this may put the business owner at risk.

PRO POINTER

The customer should be asked to sign the service order before the work is started and again when the work is completed.

Liability Waivers

Liability waivers are written releases of liability. These forms protect the contractor from being accused of an act that was nearly unavoidable. You should use liability waivers whenever you believe there may be a confrontation arising from your actions.

Written Estimates

Written estimates and quotes reduce the risk of confusion about pricing with your customers, and they also present a more professional image than an oral estimate. Giving quotes verbally can lead to many problems. To eliminate these problems, put all your estimates in writing and keep copies.

Specifications

Written specifications are another way to avoid problems with your with customers regarding the exact nature of some of the materials and equipment you plan to install in their home. The type of roofing shingles, for example, can be specified by using a copy of the manufacturer's specifications and attaching it to the contract. Kitchen cabinets, sink and trim, flooring materials, windows, doors, and light fixtures can all be adequately described by referring to the specification sheets or brochures mentioned in the contract and attached to your copy and the buyer's copy of the contract.

With small jobs, the specifications can be included in the contract. When you are embarking on a large job, the specifications will generally be too expansive to put in the contract. In these instances, make reference to the specifications in the contract and attach the specification sheets to the contract, making them a part of the contract. When you are developing specifications for a job, be as detailed as possible. Include model numbers, makes, colors, sizes, brand names, and any other suitable description of the labor and materials you will be providing. Once you have a good spec list, have the customer review and sign it.

PRO POINTER

Always have the customer sign the specifications list. Without a signature of acceptance from the customer, your specs are little more than a guide for you to work from. Unsigned specification sheets carry little weight in a legal battle.

Credit Applications

All customers you are planning to extend credit to should fill out credit applications. These forms will allow you to check into the individual's past credit history. Just because a person has good references now doesn't mean you are guaranteed of being paid, but your odds for collecting the money owed are better.

Your Company Name
Your Company Address
Your Company Phone and Fax Numbers

WORK ESTIMATE

Date: ____________________

Customer name: ____________________

Customer address: ____________________

Customer phone number(s): ____________________

DESCRIPTION OF WORK

Your Company Name will supply all labor and material for the following work:

____

____

____

PAYMENT FOR WORK

Estimated price: ____________ ($ ____________)

Payable as follows: ____________________

____

If you have any questions, please don't hesitate to call. Upon acceptance, a formal contract will be issued.
Respectfully submitted,

Your Name ____________________

Title ____________________

FIGURE 9.9 Example of a work estimate form.

Cost Projections

Item/Phase	*Labor*	*Material*	*Total*
Plans			
Specifications			
Permits			
Trash container deposit			
Trash container delivery			
Demolition			
Dump fees			
Rough plumbing			
Rough electrical			
Rough heating/ac			
Subfloor			
Insulation			
Drywall			
Ceramic tile			
Linen closet			
Baseboard trim			
Window trim			
Door trim			
Paint/wallpaper			
Underlayment			
Finish floor covering			
Linen closet shelves			
Closet door & hardware			
Main door hardware			
Wall cabinets			
Base cabinets			
Countertops			
Plumbing fixtures			
Trim plumbing material			
Final plumbing			
Shower enclosure			
Subtotal			

FIGURE 9.10 Example of a cost projections form *(continued on next page).*

Item/Phase	*Labor*	*Material*	*Total*
Light fixtures			
Trim electrical material			
Final electrical			
Trim heating/ac material			
Final heating/ac			
Bathroom accessories			
Clean up			
Trash container removal			
Window treatments			
Personal touches			
Financing expenses			
Miscellaneous expenses			
Unexpected expenses			
Margin of error			
Subtotal from first page			
Total estimated expense			

FIGURE 9.10 *(continued)* Example of a cost projections form.

As a business owner, you can subscribe to the services of a credit-reporting bureau. Normally, for a small monthly fee and an inexpensive per-inquiry fee, you can pull a detailed credit history on your customers. You will, of course, need the permission of your customers to check into their credit background. Credit applications provide documentation of this permission.

Even if you don't belong to a credit bureau, you can call references given by the customer on the credit application. This type of investigation is not as good as the reports you receive from credit agencies, but it is better than nothing. Credit applications should be used for all your customers wishing to establish a credit account.

Good Contracts Make Satisfied Customers

When the home project is in its beginning stages, most customers are excited and enthused, but as the job progresses many of these same happy customers become disgruntled. You need satisfied customers to get return business and referrals. How can you keep your customers happy? Easy-to-understand contracts will go a long way in keeping consumers appeased. A majority of consumer complaints arise because of

Option	Vendor	L/M	Price	Notes
Solid trim				
Stained trim				
Solid doors				
Stained doors				
Valance over kitchen cabinets				
Dishwasher				
Garbage disposer				
Automatic garage door				
Opener				
Tile work				
Wood floors				
Fancy handrails				
Wood stairs				
Security system				
Cable television pre-wire				
Telephone pre-wire				
Intercom				
Additional oil tank				
Additional attic/crawl lighting				
Ground cover in crawl				
Mirrors				
Plywood instead of wafer board				
2 layers of subfloor or ¾ T & G				
Insulated wall sheathing				
Ridge vent				
Soffit vents				
Gable vents				

FIGURE 9.11 Example of an estimating form for options *(continued on next page)*.

Option	Vendor	L/M	Price	Notes
Gutters				
Rain diverters				
Shutters				
Window screens				
Garage				
Deck				
Fireplace				
Power venter				
Flue				
Domestic coil				
Hot-water tank				
Cleaning				
Trash removal				
Landscaping				
Walkways				
Porches				
Vented range hood				
Tub/shower doors				
Sump pump & piping				
Flood lights				
Ceiling lights				
Ceiling fans				
Overhead/underground electrical service				
Wall wrap				
Rigid foam insulation				
Pull-down attic stairs				

FIGURE 9.11 ***(continued)*** Example of an estimating form for options.

Job Name: ____________________

Phase: Foundation

Contractor: ____________________

All Work To Be Done According To Attached Specifications

Bid Item

Supply labor and material for footings
Supply labor and material for foundation walls and piers
Supply and install foundation windows/vents
Supply labor and material to create bulkhead opening, ready for door installation
Supply and install foundation bolts
Remove all foundation clips
Waterproof foundation to finished grade level
Supply labor and material to install concrete basement floor

FIGURE 9.12 Foundation bid sheet.

Job Name: ____________________

Phase: Paint

Contractor: ____________________

All Work To Be Done According To Attached Specifications

Bid Item

Provide price for labor and material to paint, stain, and/or seal all surfaces specified
Price should include all preparation work required (i.e., filling nail holes)

FIGURE 9.13 Painting bid sheet.

Job Name: ________________________________

Phase: Drywall

Contractor: ________________________________

All Work To Be Done According To Attached Specifications

Bid Item

Supply and install all materials needed to drywall all interior walls and ceiling to code requirements

Provide separate labor only price for hanging, taping, and finishing drywall

If heat is needed, drywall contractor shall supply it

Provide separate price for texturing ceilings

FIGURE 9.14 Drywall bid sheet.

Job Name: ________________________________

Phase: Heating

Contractor: ________________________________

All Work To Be Done According To Attached Specifications

Bid Item

Supply and install all rough heating materials and finished heating equipment including boiler and baseboard units

FIGURE 9.15 Heating bid sheet.

Job Name: ____________________

Phase: Tree Clearing

Contractor: ____________________

All Work To Be Done According To Attached Specifications

Bid Item

Cut all trees marked with blue ribbons
Remove all wood, branches, brush, and debris from cutting procedure.

FIGURE 9.16 Tree clearing bid sheet.

Job Name: ____________________

Phase: Plumbing

Contractor: ____________________

All Work To Be Done According To Attached Specifications

Bid Item

Supply and install all rough plumbing and plumbing fixtures, including bathing units
Bid a separate price for well pump and related equipment

FIGURE 9.17 Plumbing bid sheet.

Job Name: ____________________

Phase: Well

Contractor: ____________________

All Work To Be Done According To Attached Specifications

Bid Item

Supply labor and material to install drilled well with steel casing and cap
Supply labor and material to install submersible pump and related equipment
Bid job on a per-foot basis and on a flat-fee basis

FIGURE 9.18 Well bid sheet.

Job Name: ____________________

Phase: Siding

Contractor: ____________________

All Work To Be Done According To Attached Specifications

Bid Item

Install siding materials provided by general contractor

FIGURE 9.19 Siding bid sheet.

Job Name: ____________________

Phase: Framing

Contractor: ____________________

All Work To Be Done According To Attached Specifications

Bid Item

Supply labor to frame house to a dried-in condition
If a crane is needed, it will be at the framing contractor's expense
Install all exterior windows and doors
Subfloors are to be glued and nailed
Provide access for bathtubs and showers
Install ceiling strapping
Install all steel beams and plates as might be required
Install all support columns
Build stairs during initial framing

FIGURE 9.20 Framing bid sheet.

Job Name: ____________________

Phase: Flooring

Contractor: ____________________

All Work To Be Done According To Attached Specifications

Bid Item

Provide price for supplying and installing underlayment
Provide price for labor and material to prepare all floor surfaces
Provide price to supply and install flooring as specified

FIGURE 9.21 Flooring bid sheet.

Job Name: ______________________________

Phase: Insulation

Contractor: ______________________________

All Work To Be Done According To Attached Specifications

Bid Item

Supply and install all insulation

FIGURE 9.22 Insulation bid sheet.

Job Name: ______________________________

Phase: Trim

Contractor: ______________________________

All Work To Be Done According To Attached Specifications

Bid Item

Supply labor to install trim materials supplied by general contractor
Provide separate price for installing counters and cabinets in all areas
Trim price should include hanging all interior doors, installing window and door hardware, and bath accessories

FIGURE 9.23 Trim bid sheet.

Job Name: ______________________________

Phase: Roofing

Contractor: ______________________________

All Work To Be Done According To Attached Specifications

Bid Item

Install roofing materials provided by general contractor

FIGURE 9.24 Roofing bid sheet.

Job Name: ______________________________

Phase: Electrical Work

Contractor: ______________________________

All Work To Be Done According To Attached Specifications

Bid Item

Supply and install temporary power pole
Supply and install all rough wiring
Install light fixtures supplied by general contractor
Electrical contractor to provide GFI devices and smoke detectors
Supply all needed permits and inspections

FIGURE 9.25 Electrical bid sheet.

Job Name: ______________________________

Phase: Site Work

Contractor: ______________________________

All Work To Be Done According To Attached Specifications

Bid Item

Remove all tree stumps and debris from any excavation
Supply and install metal culvert pipe for driveway
Install driveway—site contractor to furnish all materials
Dig foundation hole
Provide rough grading
Backfill foundation
Perform final grading
Seed and straw lawn
Install septic system
Supply and install foundation drainage
Supply and install crushed stone for foundation
Dig trenches for water service and sewer
Backfill trenches for water service and sewer

FIGURE 9.26 Site work bid sheet.

poor communication and confusion. When you remove the risk of confusion, you improve the odds of ending the job with a happy customer.

When you prepare contracts, put as much information in them as possible. Once the contract is written, go over it with the customer. Allow the customer plenty of time to read and absorb the contents of the contract. Answer any questions the customer may have, and if necessary reword the contract to eliminate confusion. Once the contract is agreeable to all parties, execute it, preferably in front of a witness.

After the contract is signed, don't deviate from its contents. If changes are to be made, use addendums or change orders, and make sure they are signed by all the signatories on the contract. If you follow this type of procedure in all your jobs, you should have more happy customers and fewer problems and you'll end up making more money.

PRO POINTER

Contracts that are worded clearly and contain all details of the job will eliminate the likelihood of confusion.

CHAPTER TEN

Trucks, Tools, Equipment, and Inventory

Equipment, vehicles, and inventory will account for most of your start-up costs, and if you are already in the business, you know that these items consume a large portion of your operating capital. There is no question that even buying a basic-model $18,000 truck can put a dent in your bank account. With a 10-percent down payment and a five-year loan at 10-percent interest, you may have to pay $1,800 as a down payment and about $344 a month. Add to this the cost of registration, taxes, insurance, and maintenance, and you have a major expense.

You may have to buy some equipment. The type and cost will depend upon your type of business. Quality hand tools are expensive and power tools more so. As for inventory, a remodeler can easily invest $10,000, but it is possible to work will very little by buying only what you need when you need it. That is a smart way to do things.

Large debts incurred without steady flows of income are the reason why large numbers of business are forced to close each year. Some business owners turn to leasing vehicles instead of buying them outright; this saves up-front money and frequently the payments are less. But if these leased vehicles rack up excessive mileage or incur major body damage, be ready to face some stiff charges at turn-in time.

Leasing vs. Purchasing

Most business owners compare leasing and purchasing tools and vehicles. Both options offer advantages and disadvantages. It makes a lot of sense to rent specialty tools that you only use a few times a year. Leasing vehicles can save you money and provide tax advantages. So what should you do? Well, let's find out.

Renting Tools

Renting expensive tools can be a very smart move for a new business. When you are first starting out, money can be scarce, and you may not know exactly what tools you need. By renting what you need when you need it, you have several advantages. You can try them out before you buy them, evaluate how often you'll need the tool, and determine how much extra money you can make with the tool.

Buying Specialty Tools

Buying specialty tools can be a mistake. If you buy a bunch of tools you don't use very often, you will be depleting your cash. For example, I used to do a high volume of basement bathrooms. Installing these basement baths required the use of a jackhammer. I started out renting the hammers and wound up buying one. For me, this was a good move, but, depending upon how much chipping you do, the outlay to purchase can't be justified. And you also have to think about the costs to maintain and repair these types of tools. Unless breaking up concrete floors is a routine part of your business, the cash outlay for a jackhammer may well be a mistake.

Before you buy specialty tools, make sure you need them. The best way to assess your needs for specialty tools is to rent them when you need them and keep track of how they affect your business. If your profits increase, consider purchasing the tool. If you find you only rent the tool a few times a year, continue to use rental tools.

PRO POINTER

When you rent, you leave the maintenance and routine repairs to the rental company, but when you own, you also own the cost to keep these tools in top working condition.

Leasing Vehicles

Leasing vehicles can be either a good or a bad idea. On the plus side: minimal out-of-pocket cash, normally lower payments, possible tax savings, and possible short-term commitments. The disadvantages are no equity gain, more concern for the condition of the vehicle, and possible cash penalties when the lease expires.

Most auto leases only require the first month's rent and an equal amount for the down payment. In other words, if the vehicle is going to cost you $250 a month, you will need $500 for a down payment and the first month's rent. To purchase the same vehicle, you might need $1000 to $1,500 for a down payment.

The monthly payments on leased vehicles are usually lower than the payments for a vehicle purchase because the truck or car will have residual value at the end of the lease. How much you save will depend on how expensive the vehicle is, but the savings will add up. However, you don't own the vehicle, and at the end of the lease, you will have no equity in the truck or car.

Leases can usually be obtained for any term, ranging from one year to five and sometimes more. Two- leases and four-year leases are popular. By leasing a vehicle for a short time, such as two years, your company fleet can be renewed frequently. This keeps you in new vehicles and presents a good image.

Most leases allow for a certain number of miles to be put on the vehicle during the term of the lease. If the mileage is higher than the allowance when the lease expires, you will have to pay a certain amount per mile for every mile over the limit. This can get expensive. Additionally, if the vehicle is damaged or abused inside or out, you can be charged for the loss in the vehicle's value. This can also amount to a substantial sum of money.

Then there are the tax angles. Since I'm not a tax expert, I recommend that you talk with someone who is. It is likely that leasing will be more beneficial to your tax situation than buying, but check it out.

PRO POINTER

If the vehicle is not in good shape at the end of the lease, you will pay a price for the abuse.

Purchasing Vehicles

Many companies lease cars and trucks, and a lot of companies buy their vehicles. When you purchase your vehicles, you are building equity in them. If you abuse the vehicle, you will lose money when you sell or trade it but you won't be penalized by a lease agreement. High mileage use or rough service usage might lead to a consideration to purchase.

Online purchasing of both new and used cars and trucks is a fairly recent phenomenon, and shopping around can be

PRO POINTER

If you need trucks that will lead a hard life, purchasing them is probably a good idea.

done in one evening to sort out the exact model and equipment of the car or truck being considered. There are services online that can tell you exactly how much a specific vehicle costs the dealer so you can go in armed with your best offer and know how to negotiate properly.

Separate Needs from Desires

Before you buy expensive equipment, vehicles, or inventory, separate your needs from your desires. When you see that fellow driving a Peterbuilt or Kenworth dump truck complete with chromed wheels, two vertical chromed exhausts, and fancy polished aluminum truck body, do you often think about how much extra work he has to perform to pay for all that glitz? If you can do the job with a $19,000 van, don't buy a $25,000 truck. There is a big difference between a need and a desire. Let's examine how you can decide what to buy.

PRO POINTER

Too many contractors get caught up in having the best and most expensive items available.

Most new business owners want to be well stocked with inventory, but they often wind up having lots of money tied up in stock that must move in order to be profitable. Before you start investing in materials and parts, figure out what you need, not what you want.

I'll give you an example of how a poorly thought-out plan can damage your business. I consulted with a plumbing company a few years ago. This particular company had eight service plumbers on the road and a ton of unneeded inventory. The business owner couldn't understand why he was short on cash. When I assessed the business, I found several faults, but inventory was one of the major problems with the company cash flow.

Being a master plumber and a business owner for many years, I know what a plumber needs on a service truck. When I looked inside this company's trucks, I was amazed. Instead of having one box of copper elbow fittings, these trucks had as many as five boxes. There are fifty three-quarter copper elbows in a box. Granted, this is a frequently used fitting, but you don't need 250 els on a service truck.

My inspection turned up countless items in multiple quantities that were unlikely to be used in a year's time. When we streamlined the trucks by reducing their inventory to a reasonable level, the company returned the excess inventory and received a credit from the supplier for over $12,000, and that was just truck stock. When we went through the back room, we eliminated several more thousands of dollars in inventory. Just by reducing the inventory to what was needed, the company generated close to $20,000 even after paying restocking fees.

Many business owners have to fight the temptation to buy more tools than they need. They sometimes seem to want every tool and piece of equipment they could ever hope to use. For these occasional-use items, rent them. A carpenter that specializes in interior trim work does not need a set of pump jacks. A builder who subs out all grading work doesn't need a bulldozer or tractor. You must decide if you simply want something or if you need it.

For years I wanted my own dump truck. I thought it would be great to own such a truck. I came close to buying one, so close that I was sitting down with the salesman. But, at the last minute, I bought a van instead. Why did I buy the van? The van would work for my needs, and it would get another crew on the road. The van would pay for itself and the dump truck wouldn't. This is the type of analysis you must make.

PRO POINTER

It is good business to have the tools and equipment you need, but it is senseless to buy expensive items that you will rarely use.

Financial Justification

Financial justification is the key to making a wise buying decision. Just like my dump-truck example, you have to see if what you are about to do is economically feasible. Justifying a purchase isn't enough; you must justify the purchase financially. There are different ways to do this.

To justify a purchase, rent long enough to establish a true need for it. Or you might use subcontractors for some part of your work until you see that it would be cost-effective to put your own people in place to take care of certain aspects of your job. For example, it might be feasible to hire your own plumber or electrician, but it probably wouldn't pay to put a cleaning crew on your payroll. Test the waters before you jump in with both feet.

How Much Inventory Should You Stock?

How much inventory should you stock? Stock just as much as you need and not a bit more. Inventory requirements vary with different types of businesses. Where a remodeling contractor might need a rolling stock worth thousands of dollars, a builder has little need for such an extensive inventory. You have to establish your inventory needs based on your customers' buying trends.

New Construction

The construction of a new house doesn't require much inventory. You can order a specific amount of material for the job when it starts. It is usually best to have some inventory on hand in case you forgot to order some everyday items in the big order or underestimated quantities. As an example, you might want a few extra boxes of nails on your truck, but you wouldn't be likely to haul around a bunch of framing lumber.

Time-Savers

An inventory of frequently used materials can be a real time-saver. If you are on a job and need an extra roll of roof flashing, you will save time and money if you have one on the truck. It is a good idea to carry a minimum quantity of your most frequently used items, but don't get carried away. Many builders don't carry excess material with them to job sites. That's okay. Plumbers and electricians who do service work need trucks with a lot of stock on them, but builders don't and remodelers don't need as much truck stock as service plumbers. Figure your jobs accurately and order what you need on a schedule that allows you to keep up with the production schedule.

Stock in Your Yard or Warehouse

Many contractors have a tendency to stock leftovers from various jobs. If you have a partial roll of insulation left over on a job, save it. However, if you ordered a range hood that is the wrong color, don't save it for another job. Return it and get credit for the item from your supplier. Money invested in inventory is money that is tied up. You should be able to find a better use for your cash.

PRO POINTER

Limit stock in your yard or in your warehouse to what you will use in a two-week period. As you use the material, order new stock. This keeps your inventory fresh and your money turning over.

Controlling Inventory Theft and Waste

Controlling inventory theft can be a problem, and it can occur in any company. Whether you have one employee or 100 employees, you need to be sensitive to the fact that some employees will take materials home with them. And material left outside on an isolated construction site is an open invitation to theft. Lost inventory is lost money. You must take steps to ensure that your employees are honest and that the material is going where it was meant to go.

The best way to reduce inventory pilfering is to keep track of your inventory on a daily basis. Have your workers fill out forms for all material used. Have the forms completed and turned in each day. Let employees know, in a nice way, that you check inventory disbursement every week. This tactic alone will greatly reduce the likelihood of employees stealing from you.

Another way to control inventory is to issue all stock yourself. If you don't allow employees access to your inventory, they can't steal it. Stock on the truck is harder to account for. However, if your employees know you keep daily records of your stock, they will be less likely to empty your truck. It never hurts to do surprise inspections and inventories on the trucks. When you do this, make sure the employees see you inspecting the trucks. The fact that you go on the truck or on a job site and audit the inventory will reduce your losses. The fewer tools left on the site, the better. There have been instances where tools stored in a heavy-duty steel tool box have disappeared and large tools secured with heavy chains have been broken by an experienced thief. Waste is another factor that can raise havoc with your inventory. Thoughtless cutting of full sheets of plywood for a small piece or cutting 6 inches or a foot off a full- length 2x4 instead of scouting around for scraps can easily eat into your inventory. Instruct your carpenters on the way to fully utilize each bit of material on site.

> **PRO POINTER**
>
> It may be better to order smaller quantities of lumber, sheetrock, pipe, and the like and pay a slightly higher premium to reduce the amount of materials stored on site.

Stocking Your Trucks Efficiently

Stocking your trucks efficiently is critical for success and maximum profits. Keep your rolling stock to a minimum; you never know when the truck will be broken into or stolen. It is more difficult to monitor employees who deal with mobile inventories. If the inventory on your trucks is collecting dust, get rid of it and don't reorder it.

A good way to establish your truck-stock needs is to keep track of the materials you use. If you track your material usage, you will be stocking your truck with materials you are very likely to use. If you have some slow movers on the truck, don't replace them when they are used. Conversely, if you have some hot items, keep them on the trucks.

Stocking your trucks efficiently will take a little time; start small and work your way up. If you have been in the trade for a while, you will have a good idea of your inventory needs. By tracking your material sales, you can perfect your rolling stock inventory.

There is a big difference in the inventory needs of a remodeler and a builder. Most builders have little need for inventory. Remodelers do often need odds and ends on their trucks. Your inventory needs will be specific to your business. As your business and experience grow, so will your knowledge of what is and isn't required.

C H A P T E R E L E V E N

Creating and Promoting an Attractive Business Image

How much is your public image worth? Well, the type of business image you present can mean the difference between success and failure. What about such things as a company logo? People quickly learn to associate a logo with its owner. Logos can do a lot for you in all your display advertising by creating familiarity between readers and your company. The name you choose for your company is important and needs careful consideration. Some names are easier to remember than others. If it is your intention, long range, to sell your business, the company name should not be too personalized. The new owner may not like owning a business with your name as part of the company name.

A company image can affect the type of customers the business attracts. Your business advantage can originate from your involvement in community organizations. How you shape your business image may set you apart from the competition. It is difficult to put a price on your public image. While it is difficult to set a monetary value on your company image, it is easy to see how a bad image will hurt your business. Your public image has many facets. Your tools, trucks, signs, advertising and uniforms will all have an impact on your corporate image. This chapter is going to detail how these and other factors work to make or break your business.

PRO POINTER

A name can conjure a mental image, an image you want for your business.

Public Perception Is Half the Battle

How the public perceives your business is half the battle. If you give your customers the impression of being a successful business, your chances of being successful will increase. On the other hand, if you don't develop a strong public image, your business may sink into obscurity.

How does the public judge your public image? There are many factors that contribute to how the public perceives your business. Take your truck as an example. What kind of image do you think an old, battered pick-up with bald tires and a license plate hanging from baling wire will project? Would you rather do business with this individual or a person driving a late model, clean van that had the company name professionally lettered on the side? Which truck points to the most company success and stability? Most people would prefer to do business with a company that gives the appearance of being financially sound. This doesn't mean you have to have flashy new trucks, but they should be well maintained.

It is important to have your company name on the business vehicles. The more people see your trucks around town, the more they will remember your name and develop a sense of confidence. It is acceptable to use magnetic signs or professional lettering, but don't letter the truck with stick-on letters in a haphazard way. Remember, you are putting your company name out there for all the world to see. You want to attract attention, but not the kind you'd like to attract.

Designing your ad for the phone directory is another major step in creating a company image. As people flip through the pages of the directory, a handsome ad may stop them in their tracks. An eye-appealing ad can get you business that would otherwise be lost to competitors.

PRO POINTER

Any professional salesperson will tell you that to be successful, you must always be in a selling mode. It doesn't matter where you are or what you're doing, you must be ready to cultivate sales and your image every day and in every way. You can't afford to let your business image slip.

While we are talking about phone directories, let's not forget about phone manners. Telephones often provide the first personal link between your business and potential customers. If you lose customers at the inquiry stage, your business will suffer. You could lose customers if you allow small children to answer your business phone. An answering service with an abrupt message is a sure way to lose potential business. Answering machines may cost you some business, too, but they are an acceptable form of doing business. Customers are calling and expect a professional response. If you put a tape in your answering machine that is non-professional or offensive, would-be customers are sure to hang up.

If your company image is strong, customers will come to you. They will see your trucks, job signs, and ads and call you. When a customer calls a contractor, they are usually serious about having work done and your company image will help you in landing the job. By building and presenting the proper image, you are halfway home to making the sale.

Picking a Company Name and Logo

Picking a company name and logo should be considered a major step in building your business image. The name and logo you choose will be with the company for many years, hopefully. Before you decide on a name or a logo, you should do some research and some thinking. There will be questions you have to ask yourself. For example, do you plan to sell the business in later years? If you do, pick a name that anyone could use comfortably. A name like Pioneer Plumbing can be used by anyone, but a name like Ron's Remodeling Services is a little more difficult for a new owner to adopt. This is only one example. Let's move on to other considerations in choosing a name and logo.

Company Names

Company names can say a lot about the business they represent. For example, High-Tech Heating Contractors might be a good name for a company specializing in new heating technologies and systems. Solar Systems Unlimited could be a good name for a company that deals with solar heating systems. Authentic Custom Capes would make a good name for a builder that specialized in building period-model Cape Cods. How would a name like Jim's Custom Homes do? It might be alright, but it doesn't say much. A better choice might be, Jim's Affordable First-Time Homes. Now the name tells customers that Jim is there for them with affordable first homes. See how a name can influence the perception of your business?

It has been proven that people remember things that they are shown in repetition. When you run an ad in the paper every week, people will remember your ad. Even though readers may not realize it, they are committing your company name or logo to their subconscious. Then, when these people scan the pages of a phone directory for a contractor and run across your name or logo, you have an edge.

Since these potential customers have been given a steady dose of your advertising, your name or logo will stick out from the crowd. Without thinking about it, people that have had regular exposure to your advertising will remember something about the ads.

Since advertising is expensive, it makes sense to get as much bang for your advertising buck as possible. If you were scanning through the newspaper and noticed a company name like Tanglewood Enterprises, what would you associate the name with? That name is not descriptive and could be used for any number of different types of businesses.

On the other hand, if you saw a name like Deck Masters, Inc., you would associate the name with decks. If the name was White Lighting Electrical Services, you would think of an electrical company. A name like Homestead Homes gives a clear impression of a company that offers warm, comfortable housing. The more you can equate the name of your business to the type of business you are in, the better off you will be. If you can add in descriptive words your customers will know more about your business, just from the name.

It also helps when choosing a name for your company to find words that flow together smoothly. How does Pioneer Plumbing sound to you? Both words start with a "P", and the words work well together. A name like Ron's Remodeling sounds good and so does a name like Mike's Masonry. In contrast, a name like Englewood Heating And Air Conditioning is not bad. A name like Septic Suckers flows well and might be fitting for a company offering service to pump out septic tanks, but the name may be offensive to some people. Your company name says a lot about your business. Maybe your company name would be something like Best Builders or Built To Perfection. Find a name that identifies what you do best.

PRO POINTER

Your company name should be one that you like, but it should also work for you. If you can imply something about your business in the name, you have an automatic advantage.

Logos

Logos can be as important as your company name. Logos, the symbols that companies adopt to represent their existence, play an important role in marketing and advertising. While people may not remember a specific ad or even a company name, they are likely to remember distinctive logos. If you put your mind to it, I'll bet you can come up with at least ten logos that stick in your mind.

Do you remember what gas company used to put a tiger in your tank? In the cola wars, who has the "Real Thing"? If you were going shopping for tires, whose ad would you remember? My guess is that you might remember the baby riding around in a certain brand of tire. Who lets you reach out and touch someone, by phone, of course? You see, it is easy to remember ads, slogans, and jingles. It is equally easy to remember logos.

PRO POINTER

Major corporations know the marketing value of logos and invest considerable time and money in coming up with just the right symbol to represent their corporation. Like slogans and jingles, logos are often much easier to remember than company names.

Your logo doesn't have to be complex. In fact, it might be nothing more than the initials of your company name. Then again, you may have a very complex logo, one that incorporates an image of what your business does. One of my favorite logos was used by a real estate company. The logo was a depiction of Noah's ark, complete with animals. In the ad featuring the ark were the words, "Looking For Land?" The business was selling land, and the ark logo was humorous and fitting for the occasion.

If it is difficult for you to create images and marketing, it may serve you well to consult with a specialist in the field. Choosing the proper name and logo is important enough to warrant investing some time and money. Of course, you know your financial limitations, but if you can afford it, get some professional advice in designing the image of your company.

How Your Image Affects Your Clientele and Fee Schedule

How does your image affect your clientele and fee schedule? Image may not be everything in business, but it is a big part of your success. People have become wary of contractors. The public has read all the horror stories of rip-offs and contractor con artists, some of which is true, and the public does have a right to be concerned. With the growing awareness of consumers, image is more important than ever before. Let's take a quick look at three examples of making a sales call.

The Visual Image

The visual image of you and your business can influence the profits of your company. In our first example, a contractor goes on an estimate in well-worn work clothes and driving a truck that has seen much better days. While this image may not offend or alienate some customers, it will surely turn many customers off.

On the other hand, you can go overboard image-wise. You drive up in a luxury car wearing a suit that cost more than the first contractor's truck. Some homeowners will relate and respond well to that image, especially if you are catering to the high end of the market.

If you dress neatly, even in casual clothing, and drive a respectable, well maintained vehicle, your odds of appealing to the masses improve. By wearing clothes that make you believable as a skilled tradesperson, you give the impression of someone who knows the contracting business. Your vehicle looks professional and successful. For me, this combination has always worked best.

The same basic principles apply to your office. If your office is little more than a hole in the wall with an answering machine, an old desk, and two broken chairs, people will be concerned about the financial stability of your business. However, if your office is staffed with several people, decorated in expensive art and furnishings, and in an expensive location, customers will assume your prices are too high. It generally works best to hit a happy medium with your office arrangements.

PRO POINTER

You need to moderate and adjust the image you are trying to project to the type of job you are selling and the type of customer you are selling it to.

Fee Factors

What does image have to do with the fee you charge for your services? To a large extent, people feel that they get what they pay for. The image you present has a direct effect on your fees.

If you convince potential customers you are a professional, the customers will be willing to pay professional fees. Extend your image by making the customers feel secure doing business with you and you have leverage for even higher fees. When you become a specialist, you may be able to demand higher fees. Think about it, who gets a higher hourly rate, your family doctor or a heart specialist?

For years I specialized in kitchen and bath remodeling. My crews did nothing but kitchen and bathroom remodeling. When you do the same type of workday in and day out, you get pretty good at it.

With my experience in this specialized field of remodeling, I could anticipate problems and find solutions before most of my competitors could. This specialized experience made me more valuable to consumers. I could snake a two-inch vent pipe up the wall from their kitchen to their attic without cutting the wall open. I could predict with accuracy how long it would take to break up and patch the concrete floor for a basement bath. In general, I became known as a competent professional in a specialized field, and I could name my own price, within reason, for my services.

PRO POINTER

When you convince the customer that nobody builds a better house than you do, you are building a case for higher fees.

PRO POINTER

If you have a special skill and can show the consumer why you are more valuable than your competitor, the consumer is very likely to pay a little extra for your expertise. Building a solid image as a professional that specializes in a certain field has its advantages.

Once You Cast an Image, It Is Difficult to Change

Once an image is cast it is difficult to change. If you are already operating an established business, it is more difficult to change an established image than it is to create a new one. If during the building of your company image you have found flaws, work to change them. With enough time, effort, and money, you can make a difference in your company image.

Let's say you started your business without much thought. You picked a name out of thin air and you never got around to designing a logo. Now you realize that you have hurt the prospects of your business getting off the ground. What should you do? You must make changes to correct your mistakes.

You can create a logo, and you can change the direction of your company, but changing the name can get tricky. If you change the name abruptly, you may lose existing customers. How will you accomplish your goal of changing the name? The procedure is not as difficult as you may think.

When you want to change the name of your existing company, do a direct-mail campaign on your existing customer base. Send out letters to all of your customers advising them of your new company name. Explain that due to growth and expansion, as an example, you are changing the name of the company to reflect your growth and your new services. Impress upon the existing customers that the company has not been sold and is not under new management, and unless you have a bad image to overcome, the new-management announcement might be a good idea.

Start running your new advertisements with the new company name and logo. Build new business under your new name and convince past customers to follow you in your expansion efforts. By taking this approach, you get a new public image without losing the bulk of your past customers. While this approach will work, it is better to take the time and effort to create a good image when you begin the business.

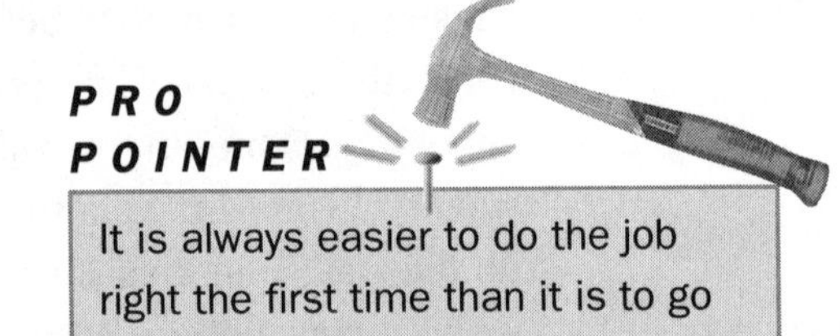

It is always easier to do the job right the first time than it is to go back and correct mistakes.

Set Yourself Apart From the Crowd

In order to make your business better than average, you must set yourself apart from the crowd. You can do this with a logo, company colors, slogans, and any other ways that are appropriate for your business. We have already talked about logos, so let's look at some other ways to give your company a unique identity.

Company Colors

Company colors are one way to attract attention and become known all over town. Well, if you don't think color makes a difference, ask the cab drivers that ride around in yellow cars. Need another example? How about the colors red, white, and blue, what do they mean to you? Colors can have a strong impact on what we think of and the context we think of it in.

There are consultants who specialize in colors. These professionals work with companies in designing colors to influence consumers. Different colors affect how people think and their mood and how they react. We've all heard about bulls charging a red flag, would they charge a green flag? You see, whether the color of the flag makes a difference is not as important has what our mental image of the color says. Since we are taught that red will make the bull mad, we tend to believe it, but I'll bet there are some bulls that would be just as happy charging a blue flag.

Look at how our culture has classified the personalities of people based upon hair color. People with red hair are said to have short fuses and high tempers. Blondes are supposed to have more fun. Obviously, the color of a person's hair doesn't make them dumb, fun, or hot tempered.

Choosing the right colors for your company is important. How seriously would you take a remodeler who pulled up in a pink van with flowers painted on it? The color and decoration of the van may have no bearing on the technical ability of the remodeler, but it does cast an immediate impression. You should choose your company colors with care.

Colors are also important in your business stationary. It would be inappropriate to use fluorescent orange for your letters and lime green for your envelopes. Certainly these colors would attract attention and be remembered, but the impression would not likely be the one you wanted to create. For most businesses tan, ivory, light blue, or off white are acceptable stationary colors.

PRO POINTER

The color of your trucks may be dictated by the color of the truck you presently own. It is more impressive to see a fleet of trucks that are uniform in color and design than it is to see a parade of trucks that include various makes and colors. A unified fleet gives a better impression.

Color is also important in your truck lettering and job-site signs. If your truck is dark blue, white letters will show up better than black letters. If the truck is white, black or blue letters would be fine. For job signs, it is important to pick a background color and a letter color that contrast well. You want the sign to be easy to read from a distance so size is also important. When you talk with your sign painter or dealer, you can review samples of how different colors work together. Now, let's examine the value of company slogans.

Slogans

Slogans are often remembered when company names are not. If you will be advertising on radio or television, slogans are especially important. Since radio and television provide audible advertising, a catchy slogan can make its mark and be remembered. When advertising in newspapers or other print ads, slogans show the readers key words to associate with your company.

Try a simple test. When I ask you the question in the next sentence, no fair peeking, think of one company as fast as you can. What pizza company delivers? If you thought of Domino's, my guess was right. Domino's has been a great success in a highly competitive business. The logo on their box is a domino. Now a lot of pizza places deliver, but if you live in an area where Domino's is available, you can't think of pizza delivery without thinking of Domino's. This fact is no accident. I'm sure the brains behind Domino's have spent huge sums of money to develop this image.

The golden arches is another example of good marketing, with its slogan that lets you have it your way. All major food franchises have established logos, slogans, colors, and more. Their marketing and advertising is expensive, but it works. Advertising only costs a lot of money when it doesn't work.

PRO POINTER

If expensive advertising works, it makes you money.

Thinking of a slogan for your business might take a while, but it's worth the effort. If you need inspiration, look around you, at other successful companies. Examine their slogans to gain ideas for yours, but never use someone else's slogan.

Build Demand for Your Services Through a Strong Image

Building business demand through a strong image can be done in several ways. One way is to build your business a little each time you serve a customer. This will build a word-of-mouth referral system. Word-of-mouth referrals are the best business you can get. But, if you don't want to wait for the results of customer recommendations, you can use advertising.

Advertising is a very powerful business tool. In skilled hands, advertising can produce fantastic sales results. Consider this, you are about to move to a new city, what real estate brokerage in the new city will you call for relocation help? I would guess you would call a brokerage where all the brokers and agents wear gold coats. There we go again with company colors, but the gold-coat brokers get a lot of visibility on TV, radio, and in print ads. Once you hear a hundred times how they are the best real estate team around, you might start to believe it.

You may not know anything about a particular brokerage, but advertising plants the seed that the gold coats mean success. If you buy into the advertising, you are likely to call these brokers. If the broker doesn't make a good personal impression, you may choose another brokerage, but at least you called the gold team. This same strategy can work for you.

When operating a contracting business, advertising alone will not get the job done. You or your company representatives will have to keep the ball rolling once you are in touch with potential customers. Talk to some professionals in the field of marketing, and I think you will be surprised at the results you can achieve.

PRO POINTER

You can build demand for your services through a strong image. People like to deal with winners. If your company has the reputation of being fair, professional, competent, and dependable, customers will seek you out.

Joining Clubs and Organizations to Generate Sales Leads

Joining clubs and organizations is an excellent way to generate sales leads. As a business owner, you must also be a salesperson. When you join local clubs and community organizations, you meet people and these people are all potential customers.

By becoming visible in your community, your business will have a better chance of survival. If you support local functions, children's sports teams and the like, you become known. You can use the local opportunities to build your business image. When citizens see your company name on the uniforms of the local kids' baseball team, they remember you. Further, they respect you for supporting the children of the community. You can take this type of approach to almost any level. After you have established a public awareness of your business, you should get busy. You can't afford to let down on your marketing and advertising needs.

PRO POINTER

Marketing and advertising may well be the most important lessons for new business owners to learn. While it is true that marketing and advertising alone will not make a business a success, they are critical elements in building a thriving business. If you don't do a good job with your marketing and advertising, you won't have a chance to show customers what you can do.

There Is No Business Without Sales

There is no business without sales. To get the opportunity to generate sales, most businesses must advertise. Without advertising, the average business will have a hard time getting customer interest. If no one knows your business exists, how will they contact you for service? Since public exposure is paramount to the success of your business, so is a strong marketing plan and effective advertising.

Too many contractors fail to see the importance of marketing and advertising. For some reason, many contractors think the public will seek them out. Let me repeat myself, if the public doesn't know you exist, they can't very well seek you out. Regardless of how good you are at what you do, you won't get much work without making people aware of your services.

To get busy and stay busy, you need regular sales. Marketing and advertising can provide you with sales leads. It will be up to you or your salespeople to convert the leads into closed sales, but you must start by getting prospects wanting what you have to offer. Advertising is the most effective way to generate leads quickly.

Marketing Is a Pivotal Point to Any Business

Marketing is a pivotal point to any business. If you have the ability to perform a good market survey, you should be able to generate a vast amount of business. Advertising is the act of putting your message in front of consumers. You can advertise in newspapers, on radio, on television, by direct mail, or in many other ways. When you track your advertising results, design your ads, develop sales strategies, and define your target market, you are exhibiting marketing skills.

Marketing is much more complicated than advertising. Advertising your business requires little more than the money needed to pay for your ads. Marketing demands an extension of the normal senses. You must be able to read between the lines and determine what the buying public wants. There are many books available on marketing. Professional seminars teach marketing techniques. Many community colleges offer courses in marketing. With enough effort and self-study, you can become very efficient with your marketing ploys. If you want to have a business with a long life, you should expend the energy to develop effective marketing skills.

PRO POINTER

Marketing is not just advertising. Marketing is reading the business climate.

Should You Enlist Commissioned Salespeople?

This is a good question, and the answer lies within what your business goals are. Commissioned salespeople can make a dramatic difference in your business. Commissioned salespeople can generate a high volume of gross sales and, since you are paying the sales staff only for what they sell, an army of sales associates can be mighty enticing. However, a high volume of sales can create numerous problems. You may not have enough help to get the jobs done on time. You might have to buy new trucks and equipment and increase your workforce quickly without having enough time to determine if they are fully qualified. The increased business may tie you to the office and cause your field supervision to suffer. There are many angles to consider before bringing a high-powered sales staff online.

The Benefits Of A Sales Staff

The benefits of a sales staff are many. If you find the right people to represent your company, you can enjoy increased sales. By having commissioned salespeople, you don't have the normal overhead of employees, and so you only pay for what you get. Good salespeople will generate deals that may otherwise never come your way. A strong deal closer will make deals happen, so you have quick sales. Sales professionals can convert a simple job and turn it into a major job for you. With the right training and experience, sales professionals can get more money for a job than the average contractor would. It is clear that for some businesses a sales staff is a powerful advantage.

The Drawbacks To Commissioned Sales People

The drawbacks to commissioned sales people may outweigh the advantages. Some salespeople will tell the customer anything they want to hear to get a signature on the contract. As the business owner, you will have to deal with this form of sales embellishment at some point during the job. The customer might tell you that the salesperson assured them that they would get screens with their replacement windows, when you had not figured screens into the cost of the job. The salesperson might have promised that the job could be done in two weeks, when in reality, the job will take four weeks. This type of sales hype can cause some serious problems for you and your workers.

Most sales associates are not tradespeople. They don't know all the technical aspects of a job; they only know how to sell the job, not how to do it. A salesperson might tell a prospect that putting a bathroom in the basement is no problem, when in fact, such an installation requires a sewer pump that adds nearly $800 to the cost of an average basement bath. There are many times when an outside sales staff undersells a job. Sometimes they sell the job cheap to get a sale. At other times the wrong price is quoted out of ignorance. In either case, you, as the business owner, have to answer to the customer.

Sending the wrong person out to represent your company can have a detrimental affect on your company image. If the salesperson is dishonest or gives the customer a hard time, your business reputation will suffer. Deciding when to use commissioned salespeople is your decision. But, let me tell you, don't take the decision lightly. There is no question that the right salespeople can make your business more profitable. However, there is also little doubt that the wrong sales staff can drive your business into the ground.

PRO POINTER

Getting too many sales too quickly can be as devastating as not having enough sales. If the salesperson you put in the field is good, you might be swamped with work. This can lead to problems in scheduling work, the quality of the work turned out, field supervision, cash-flow, and a host of other potential business killers.

If you decide to use commissioned salespeople, I suggest you go with them on the first few sales calls. When you are interviewing people to represent your company, remember they are sales professionals. These people will be selling you in the interview with the same tenacity that they will use on prospects in the field. Go into the relationship with your eyes wide open. Don't take anything for granted and check the individuals out for integrity and professionalism.

Where Should You Advertise?

Advertising in the local phone directory will generate customer inquires and provide credibility for your company. Ads in the local newspaper can result in quick responses. Door-to-door pamphlets and fliers can produce satisfactory results. Radio and television ads can be very effective, but they are expensive and require repetition to be truly effective. Putting a slide-in ad on the video boxes at the local video rental store can give you a lot of exposure. The list of possible places to advertise is limited only by your imagination. However, some advertising media are better than others. Let's take a close-up look at some specific examples.

The Phone Directory

The phone directory is an excellent place to have your company advertised. The size of your ad, however, will depend on the nature of your business and the type of work you want to attract. Being listed in the phone book ads credibility to your company.

The size of your ad in the directory should be determined by the results you hope to achieve. Large display ads are expensive, and they may not pay for themselves in your line of work. If your business is remodeling houses, a large display ad probably isn't necessary. When people are shopping for a remodeler, they are not normally in a hurry. An ad that is one column wide and an inch or two in length can create just as many calls. A quick look at how your competition advertises can give you a hint as to what you should do. If all the other contractors have large ads, you probably should have a large ad.

Over the years I have tried many experiments with directory advertising. At one time I was running a half-page ad for my business. I thought I could save money by going to a smaller ad. I did pay less for my new ad, but my business suffered from the lack of the large display ad. I had a noticeable drop in phone requests.

As I became more knowledgeable about business, marketing, and advertising, I continued to test the results of various directory ads. During my test marketing, I used many types of ads for my various businesses. I found that for remodeling, real estate, and plumbing, large ads worked best. When perfecting my ads for home building, I did just as well with smaller ads. The results of ad sizes have varied geographically for me. My requirements in Virginia called for a bigger ad than in Maine.

PRO POINTER

Whether you are merely listed in a line listing or have a full-page display ad, you should get your company name in the phone book as soon as possible.

Newspaper Ads

Newspaper ads provide quick results. You either get calls or you don't. As a service contractor, my experience has shown that most respondents to newspaper ads are looking for a bargain. If you want to command high prices, I don't think newspapers are the place to advertise. But, if you are new in business, the newspaper can produce customers for you quickly.

Handouts, Flyers, And Pamphlets

Handouts, flyers, and pamphlets are similar to newspaper advertising. These methods seem to generate calls quickly, but the callers are usually looking for a low price on your services. Many businesses consider this form of advertising as degrading. I don't know that I would agree with that opinion, but I don't think you will receive the money you are worth with these low-cost advertising methods.

Radio Advertising

Radio advertising is expensive, but it is a good way to get your name to listeners. I believe the key to radio advertising is repetition. If you can't afford to sustain a regular ad on the radio, I would advise against using this form or advertising. Most people are not going to hear your ad and run to the nearest phone to call you. However, if you can budget enough money for several radio spots for a few weeks time, you will gain name recognition.

Television Commercials

Television commercials can be very effective. People associate television advertisers with success. With the many cable channels available, television advertising can be an affordable and effective way to get your message out to the community. I have used ads on cable television very effectively. Television ads can increase your sales.

Direct-Mail Advertising

Direct-mail advertising can be very effective, but it is not always cost effective. The cost for direct-mail advertising can easily run into thousands of dollars. Most people that use this form of advertising are content if only one percent of the people they mail to become customers.

By using direct mail you can reach a targeted market. If you want to advertise to people with incomes in excess of $50,000, you can purchase a mailing list of just those people. This type of demographic break-down is very effective in mailing to the best prospects.

Most mailing lists are available at prices of around $75 for each thousand names. Many sellers of mailing lists require a minimum order of 3,000 names. The names can be supplied to you on stick-on labels. Expect extra charges for various demographic breakdowns.

If you want to reduce your mailing costs, your local postmaster can provide a bulk-rate permit. To use the bulk-rate service, you must mail a minimum of 200 pieces of mail at a time. The cost for this type of mailing is much less than first-class postage, but there are one-time and annual up-front fees to be paid. You can talk with your local post office for full details.

Creative Advertising Methods

Creative advertising methods are just that, creative. You might want to rent space on a billboard to advertise your business. Perhaps you will arrange a deal with a local restaurant to have your company highlighted on their menus. Providing uniforms for the local Little League can get your name in front of a large audience. If you put your mind to it, there is almost no end to the possibilities for creative advertising.

What Rate Of Return Will You Receive On Advertising Costs?

The response to your advertising will depend on your marketing plan and the execution of your advertising. If you are advertising in the local newspaper, you might expect about a .001 response. In other words, if the paper has 25,000 subscribers, you might get twenty-five responses to your ad. This projection is aggressive and in most cases your response will be much lower. If you advertise at the right time of the year with the right ad, twenty-five calls could come in. However, if you only get ten calls, don't be surprised. In some cases you may not even get ten calls. Your advertising success will depend entirely on how well you picked the publication and designed the ad.

PRO POINTER

Advertising a contracting business on the radio or television can seem like a waste of money. It is not uncommon for these ads to run without getting calls, but that doesn't mean the ads were not effective. Television and radio advertising builds name recognition for your company. This form of advertising works best when it is used in conjunction with some type of print advertising.

If you are running ads in the paper, distributing flyers, or doing a direct-mail campaign when the television and radio ads are on, you should see a higher response than you would without the radio and television ads.

Direct-mail advertising often provides fast results. Many people receiving ads by mail either trash them or act on them quickly. A one-percent response on direct-mail advertising is generally considered good. For example, if you mail to 1,000 houses, you should be happy if you get ten responses. Due to the low response rate of bulk mailings, direct-mail is not effective for low-priced services. However, if you are selling big-ticket items, direct mail can work very well.

If you target your direct-mail market, you should do much better on your rate of return. For example, if you use a mixed mailing list for your advertising, you don't know what type of person is receiving your ad. But, if you pick a list based on demographics, you can be sure you are reaching the type of potential customer you want.

PRO POINTER

Demographics are statistics that tell you facts about the names on your mailing list. You can rent a mailing list that consists of specific age groups, incomes, and so forth. These statistics can make a big difference in the effectiveness of your advertising.

Determining the effectiveness of advertising is a task all serious business owners must undertake. To learn what ads are paying for themselves, you need to know which ads are generating buying customers. Some ads generate a high volume of inquiries, but don't result in many sales. Other ads produce less curiosity calls and more buying customers. You need to track the results of your advertising. Without knowing which ads and advertising mediums are working, you have no way of maximizing the return on your advertising expenses.

Learn To Use Advertising For Multiple Purposes

Most businesses learn to use advertising for multiple purposes. The primary use of advertising is to generate consumer interest in goods and services. But, advertising can do much more for a business. As we have already seen, advertising can build name recognition for your company.

PRO POINTER

Advertising can be used to build your company image. Through advertising, you can create almost any look you like for your business. A company image can be responsible for commanding higher fees and quality customers.

Name recognition is important when trying to get the most mileage out of your advertising budget.

Advertising can enable you to establish your goals. If you want to be known as an expert in restoring old homes or building authentic reproductions, advertising can get the job done. As you go along in business, you will find that various forms of advertising can help you achieve success in many ways.

Building Name Recognition Through Advertising

We have already talked about building name recognition through advertising, but now we are going to learn how it's done. You want people to see or hear your company name and feel like they know the company. To accomplish this goal, you must use repetitive advertising.

Repetitive advertising can be used in all formats of advertising. Take radio advertising for example, when you hear radio commercials, you normally hear the company name more than once. Pay attention the next time you hear ads on the radio. You will probably hear the company name or the name of the product being sold at least three times.

Television uses verbal and visual repetition to ingrain a name or product in your mind. Watch a few television commercials and you will see what I mean. During the commercials you will see or hear the company name or product several times.

Not only should your name be used often in the ad, the ad should be run regularly. If you advertise in the newspaper, don't run one ad and stop. Run the same ad several different times. Use your logo in the ad, and keep the ads coming on a regular schedule. This type of repetition will implant your company name into the subconscious of potential customers. When these potential customers are ready to become customers, they will think of your company.

Generating Direct Sale Activity with Advertising

The need for generating direct sale activity with advertising is the reason most people use advertising. For a service business, generating direct sales is possible with direct mail, radio, television, print ads, telemarketing, and other forms of creative marketing. Telemarketing and direct mail are two of the fastest ways to generate sales activity.

We've already talked about how direct mail works, but how about telemarketing? Telemarketing is a tough job. Calling people you don't know and asking them to use your services, buy your product, or allow you into their home for a free inspection, estimate, or whatever is not much fun. However, if you can live with rejection and are not afraid to call 100 people to get ten sales appointments, cold calling will work.

When your objective is to generate sales activity, it helps to make your offer on a time-restricted basis. By this, I mean, offer a discount for a limited time only. Create a situation where people must act now to benefit from your advertising. Time-sensitive ads can generate activity quickly.

PRO POINTER

With new regulations on the "Do Not Call" list you will have to be very careful not to call the wrong people. Due to the risks, I have dropped telemarketing from my sales approach.

Without Advertising, The Public Will Not Know You Exist

Without advertising, the public will not know you exist. Advertising is expensive, but it is also a necessary part of doing business. If you don't spend money on advertising, the public is not going to spend more with your business. The contracting field is filled with business owners who are aggressive. These aggressive owners advertise regularly. If you don't put your name in front of people, you will be run over by the companies that do.

Promotional Activities

Promotional activities are an excellent way to get more sales and to build name recognition. By using special promotions, you capture public attention and create an opportunity for additional sales. Let me give you an example of how you could stage a promotional event.

For this example, assume you are a contractor who specializes in remodeling. You could talk to your local material supplier and develop a seminar. Have the material supplier allow you to come into the store and give a home-improvement seminar to shoppers. Tell the supplier how the seminar will be good for the store's image and increased material sales. Advertise the free seminar for about two weeks prior to the date of your talk. The supplier may be willing to pay a portion of the ad costs, after all, the store is gaining publicity from this promotion as well.

When people begin gathering around you in the store, be sure to have your business cards, rate sheets, and other sales aids displayed where the shoppers can see them. After your seminar, field questions from the audience. This type of promotion can create the image of you being an expert in your field.

If it is legal in your area, give away a door prize. Have the audience fill out cards with their names, addresses, and phone numbers for a prize drawing. Give away the prize, it could be a discount on remodeling services, a small appliance, or just about anything else you can think of. After the seminar, you have a box full of names and addresses to follow up on for work. This type of idea can increase your business dramatically.

How To Stay Busy in Slow Times

Every business owner wants to know how to stay busy in slow times. Until you have survived a business recession, you may not have the experience to stay afloat in troubled waters. Since you can't always learn survival skills on a first-hand basis and survive, you must turn to the experience of others for your training.

When times are tough, you may have to alter your business procedures. But, how will you do it? Will you lower your prices? Would you eliminate overhead expenses? How about cutting back on your advertising expenses, would you cut back on your ad budget? Any of these options could be the wrong thing to do.

PRO POINTER

If you lower your prices, you will have a very hard time working your prices back up to where they were. Lowering prices is risky business, but sometimes it is the only way to keep food on the table. If you have to lower prices to stay in business, do so with the understanding that getting prices back to normal will take time.

Discounts might accomplish the same goal as lowering labor rates, but with less long-term effects. People expect discount offers to end. Run ads offering a discount from your regular labor rates for a limited-time only. This tactic will be less difficult to deal with than a lowering of labor rates.

Cutting unnecessary overhead expenses is a good idea at any time, but cutting needed overhead expenses can be dangerous. Even when times are rough, some overhead expenses should be maintained. If you cut back on the wrong expenses, your business may get worse. For example, if you eliminate your answering service, you may miss some of the important business calls your company needs to survive. Evaluate your expenses carefully before making cuts. Only eliminate expenses that should not affect the volume and profitability of your business.

Advertising is one of the first expenditures many business owners cut. Advertising is just as vital to your business in bad times as it is in good times. In fact, advertising may be more valuable in tough times. Think long and hard before you eliminate your ad in the phone book or reduce your normal advertising practices.

Business promotion should be one of your top priorities. As I've said before, without customers, you have no business. To get customers, you must always strive to find new people to build homes for. If you get too comfortable and slack off on your business promotion, you may find that your business will shrivel up and blow away.

PRO POINTER

As your competitors cut back and shrink into obscurity, you can lunge forward with aggressive advertising.

C H A P T E R   T W E L V E

Hiring Employees

Hiring and managing employees is a daunting task for all business owners. Finding the proper balance whether to hire an employee or not may cause you to lose sleep at night. If you plan to hire an employee anticipating that business will pick up and it doesn't, you have added to your overhead. If you get increased business and you are short staffed, it may take some time to find the right employee, and in the meantime you are working 24/7. So, how will you decide how to handle the employee issue? This chapter is going to open up the questions and show you the answers. You will learn how to find, hire, manage and, if required, terminate an employee. Well, let's get on with it.

Do You Need Employees?

First of all, ask yourself why you want employees? Ah, now there's a question that requires a little thought. Do you want employees in order to make more money? Do you want employees to cast a brighter public image? Would you like to have employees so that you can feel more important? Will having employees make you more successful in your own mind? Well, let's look at each of these questions and evaluate your answers.

Why Do You Want Employees?

Many business owners can't provide a clear answer to this question. The owners may be quick to say they want employees, but they often can't express why they do.

Before you begin assembling a list of employees to expand your business building a list of employees, decide why you want to hire them. Look at your reasons and be sure they make sense. It is easy to hire employees, but it may not be so easy to get rid of them. Don't hire people until you need them and understand the reasons why.

Do You Want Employees in Order to Make More Money?

Most employers will give making more money as their reason for hiring employees. The more employees you have, the more money you will make. While this theory can hold true, it can also be dead wrong. From my personal experience, I have seen both sides of this coin. There have been times when I had several employees and made less profit than I did without them. Then, there have been occasions when employees have been good for my financial health. There were many factors in my personal experiences.

The economy has often impacted my business endeavors and affected my experience with employees. In my early years, poor management contributed to many of the problems I had with my employees. My selection of employees at various times contributed to not only my successes, but also to my failures. As we move through the chapter you will see specific examples of how employees can affect your business, both positively and negatively.

PRO POINTER

There are many factors that will influence your experiences with employees, but don't get the idea that more is always better. A lot of contractors do much better financially with small crews than they do with large crews.

Do You Want Employees to Cast a Brighter Public Image?

The public often associates the success of a company with the number of employees it has. But, is this reason enough to hire employees? No, you must hire employees for profitable reasons, not for public opinion. Public image has an affect on the success of your business, but don't burden yourself with employees for only this one reason.

Do You Need to Have Employees to Make You Feel More Important?

Most people won't admit to this, but many of them do feel more important when they have employees. This is not a good reason to hire people. There are less costly ways to improve your self-appreciation.

Will You Feel More Successful When You Have Employees?

People measure success with different measuring sticks. Some people consider themselves successful when they have a lot of money. Having a number of employees can spell success for some individuals. Family health and happiness is a common measurement of success for business owners.

How you measure success is up to you, but don't put too much emphasis on the number of employees in your business to find satisfaction. Simply having a large number of people on your payroll doesn't mean you will be happy. Unless you want to provide jobs for others while being satisfied with meager profits, employees can be a mistake.

Does It Sound Like I'm Against Hiring Employees?

I can see where you might be forming that opinion, but don't judge me too quickly. I'm not against hiring employees, but I do want to show you both sides of the issue. Many business owners never take the time to consider the downside of hiring employees. Since I assume you have dozens of reasons why hiring employees is great, I want to expose you to some of the possibilities you may not have thought of.

Do You Need Employees to Meet Your Goals?

The answer to this question may lie in an analysis of your business goals. There are limitations to what any one individual can do without help. However, you may be able to accomplish your goals with independent contractors, eliminating the need for regular employees. Let's look at the pros and cons of employees versus independent contractors.

PRO POINTER

Independent contractors may seem to be more expensive than employees when their rates are first reviewed, but further investigation may prove the independent contractors to be less expensive.

Hourly Rates

The hourly rates of independent contractors will normally be higher then the wages you would pay an employee to perform the same function. But while the hourly rate of independents is higher, they may actually cost less.

Independent contractors have many other customers and you pay them their hourly rates only when you need them. An employee, on the other hand, is on your weekly payroll and looks to you for a full week's pay. And if there are lulls in work during the week, you still have to pay your employee their full wages even though they may have been unproductive part of that time.

There are lots of hidden costs that make up an employee's total hourly or weekly cost. When subcontractors bill you on an hourly basis, they will have included in this hourly rate not only their employees' hourly wage but a whole series of other payroll related costs: insurance, workers compensation, FICA, fringe benefits such as sick pay, vacation pay, bonuses and so forth. These wage "add-ons" are referred to as the labor "burden" and in some cases adds 40% or more to the basic wage rate. Let's look at some of these labor burden items.

Insurance Costs

Independent contractors pay for their own insurance costs and these costs are included in the hourly rates they charge. These costs can be substantial. Some union ironworkers' insurance costs, computed on an hourly basis, can add $50.00 per hour or more to their base wage rate. There is liability insurance, health insurance, dental insurance, worker's comp insurance, and disability insurance that you may have to pay for with employees as long as they are on your payroll, whether they are productive or not. Although subcontractors have these same costs, once again, they will include them in their hourly rate, but only for those hours actually spent working on your job.

Transportation

Subcontractors are responsible for their own transportation. If you hire employees, you will probably have to furnish company vehicles if they are field personnel. When you consider the cost of acquiring, insuring, and maintaining vehicles, the subcontractors who provide their own transportation are desirable options.

Payroll

Payroll for companies with many employees can be a full-time job. If you add a payroll clerk, you have added another overhead expense to be recovered in the prices of your services. By using subcontractors, you eliminate the need for payroll and payroll records. Of course, you will still have to write checks to the subs, but this procedure is less labor intensive than doing payroll.

Payroll Taxes

Payroll taxes are another expense you eliminate by using subcontractors. When you have employees, you will have additional tax deposits to make for the payroll. This tax is nonexistent when independent contractors are utilized.

Paid Vacations

Most employees expect to receive paid vacations. If you give the average tradesperson a two-week paid vacation, this may cost you $1,000 or more per year. This cost applies to every employee you offer the benefit to. If you have ten plumbers, it will cost you about $10,000 a year by being a considerate boss. Subcontractors don't expect you to give them a paid vacation. This type of savings can add up quickly.

Sick Leave and Other Benefits

Sick leave and other benefits are also costs to be factored into your employee benefit package. Most employees expect these favors, but subcontractors don't.

Convenience

If convenience is a factor, employees may have an edge against subcontractors. Subcontractors are not always available when you need them. Remember, they work for many other customers and are not on call specifically for you. If you want your own people at your fingertips, employees are more readily available than subs.

Competition

Many contractors have concerns that their subcontractors will become direct competition. While some subcontractors will attempt to steal your clients, most won't. While employees are not generally looked upon as competition, they may pose more of a threat than subcontractors. Subcontractors are already in business and already have customers. Employees may be looking to go into business for themselves. They may also be considering taking some of your customers with them. From the competition angle, either group is a possible threat, but I would be more concerned about the employees.

Your decision to employ subcontractors rather than hire or increase your payroll depends upon a number of factors as we have discussed. So consider your choices wisely. But, hopefully, the one thing you have learned in this discussion is that your employee will cost you considerably more than just their hourly rate. So when you are figuring out your hourly billing rate, be sure to include all costs associated with that rate. Here is a good check list:

- Hourly wage
- Vacation pay
- Sick pay
- Holiday pay
- Car or truck allowance
- Bonus
- Payroll taxes: FICA, Medicare, Federal unemployment, State unemployment
- Health and dental insurance
- Accidental death/dismemberment insurance
- Long term disability insurance
- Life insurance
- Worker's compensation
- Travel allowance
- Uniform allowance or maintenance costs
- Union benefits (if applicable)

How to Find Good Employees

Do you know how to find good employees? If you do, you can probably make more money as a consultant to other businesses than you can in the contracting business. Business owners of all types struggle to find good employees. It is easy to hire people, but locating good employees is difficult.

Before you consider hiring employees, you must understand what makes a good employee. Defining the qualities you want in your staff should be easy. For most business owners, establishing the criteria for a good employee will not be a problem. The problem will come in trying to find these employees. Most good employees are well taken care of in their present position and won't want to leave.

Classified Ads

Classified ads are one way of finding employees, but be prepared for many wasted hours in your search. When you run a help-wanted ad, you will get responses from all types of people. Most of these people will not fit the mold you have cast. But sometimes a classified ad can produce top-notch people. If you start your employee search with the knowledge you will have to sift through a mass of unqualified applicants to find one good worker, classified ads can be worthwhile.

PRO POINTER

Classified ads are the quickest way to get applicants in your office, but they generally are not the best way to find good candidates.

Employment Agencies

Employment agencies deal mainly with executives and professionals. These agencies screen prospective applicants before divulging their client's name. Generally, the agency will contact the business owner and review the background of a prospective employee before the employer is identified. If you, the employer is interested in talking with the applicant, the agency will arrange an interview.

PRO POINTER

Employment agencies are geared more towards white-collar positions than they are to blue-collar jobs. If you are looking for people in the trades, agencies may not be of much help.

Then, there are the fees charged by agencies. Some agencies charge applicants when they locate a job for them. However, most agencies charge the employer for these fees which can be as much as 30% of the employee's annual salary. So if you were offering a job that pays $30,000 per year, the agency would bill you, in this case, $9,000 for finding and hiring that employee. Before you deal with an agency, be sure of what you are getting into. If you are asked to sign a contract or engagement letter, read it carefully and consider consulting your attorney before signing the document.

Temporary Employment Agencies

At one point, these temporary employment agencies provided only professional help, clerical, bookkeeping and accounting, and the like. But now the services offered by some of these temp agencies extend to engineers, paralegals, sales and marketing people and, you name it. There are also companies that supply temporary labor for the construction trades. This can prove to be a good source for not only part time but also for full time help. You may hire a laborer for a part time position only to find out that they are terrific at a variety of other jobs, so you might offer them a full-time job. The agency might not be too happy, but it is done a lot.

Word-of-Mouth

Word-of-mouth referrals for job applicants are very effective. If you put the word out that you are looking for help, you may find some applicants through your existing employees or friends. When a person applies for a job because a friend has recommended your company, you have the advantage of built-in credibility.

Unemployment Office Listings

Unemployment offices carry listings of job opportunities for people in need of work. If you have a job to fill, notify the local unemployment office. They will put your job listing in their computers and on their bulletin boards. This type of listing service is free and can produce quick results.

Hand Picking

PRO POINTER

It is frowned upon to steal employees from your competition.

Hand picking employees is one of the best ways to get what you want. However, you must remain ethical in your approach. Let's look at how you should and shouldn't go after specific employees.

There is a right way and a wrong way to hand pick employees. First, let's look at the wrong way. Years ago, I was a project superintendent on a townhouse project. I was in charge of all the plumbers and support people for the plumbers. During these times, the economy was good and good plumbers had jobs. However, most plumbing companies needed more plumbers to keep up with the rapid building trends. There were companies that showed no remorse in stealing good plumbers from the competition. During this project, I saw competitors come onto my job and offer my plumbers more money, right in front of me!

The plumbers knew they could get work anywhere, and many of them would change jobs for an extra twenty-five cents an hour. When these people raided my job, they often left with my men in their trucks. The plumbers didn't give any notice; they just left, following the higher hourly wages. Some plumbers even tried to get me and the other company's representative into bidding wars.

You can imagine the ill will that formed between companies under these circumstances. One company would steal a plumber on Monday, and on Friday, another company would take the plumber away to a new job. This endless turnover of plumbers hurt everyone in the business. You don't want to use these techniques in hand picking employees.

What tactics can you use to hand pick employees? There are many ways to make employees aware that you are interested in offering them a position. One excellent way is to run into them at the supply house. While the two of you are standing around, waiting for your orders to be filled, start a conversation that leads to your need for help. Stress how you are looking for someone just like the person you are talking to. If the other person is interested in pursuing employment with your company, you should get some signals.

If you have your eye on a particular person for your position, don't hesitate to call or write the individual. If you don't want to look too obvious, ask the person if he knows of anyone with qualifications like his that is looking for work. By using this approach, you can give all the details you want about the position, without making a direct solicitation of the individual. There are many tactful ways to get your point across to people that are presently employed. Be discreet and keep your dealings fair.

Employee Paperwork

Let me give you a few employee pointers to provide a starting point for obeying the law in your employee paperwork. Don't take these pointers as the last word and don't consider them conclusive. Look upon them as a guide to the questions to ask your attorney and tax professional. Remember, you are trying to keep as much of your cash as you can, and losing a portion of it to fines and lawsuits is not a good use of your resources.

Employment Applications

You should have all prospective employees complete an approved employment application. These applications tell you something about the people you are considering hiring, and they provide a physical record for your files. Be sure the application forms you use are legal and don't ask questions you are prohibited from asking.

W-2 Forms

W-2 forms are used to notify employees of what their earnings were for the past year and how much money was withheld for taxes. The forms must be mailed or given to employees no later than the last day of January, following the taxable year.

W-4 Forms

W-4 forms are government forms that should be completed and signed by all employees. These forms tell the employer how much tax to withhold from an employee's paycheck. Once the form is filled out and signed, keep it in the employee's employment file. W-4 forms should be completed and signed by employees each year.

I-9 Forms

I-9 forms are employment-eligibility-verification forms. These forms must be completed by all employees hired after November 7, 1987. I-9 forms must be completed within the first three days of employment.

Employers are required to verify an employee's identity and employment eligibility. Employees must provide the employer with documents to substantiate these facts. Some acceptable forms of identification are birth certificates, driver's licenses, and U. S. passports. If an employer fails to comply with the I-9 requirements, stiff penalties may result.

1099 Forms

1099 Forms are used to report money you pay to subcontractors. These forms must be completed and mailed to them at the end of each year. You will send one copy to the subcontractor, one copy to the tax authorities, and retain a copy for your files. Since subcontractors come and go, they can be difficult to locate when the time comes to send out the 1099 forms. Insist on having current addresses on all of your subcontractors, at all times.

Employee Tax Withholdings

Employee tax withholdings cannot be ignored. When you do payroll, you must withhold the proper taxes from an employee's check. The Internal Revenue Service will provide you with a guide to explain how to figure the income tax withholdings. In addition to withholding for income taxes, you will also have to deduct for Social Security. Again, you can use a tax guidebook, available from the IRS, to figure the deductions.

Once you have computed the income and FICA withholdings, deduct them from the gross amount due the employee for wages. After doing this, enter the amount of withholdings in your bookkeeping records. While the money withheld is still in your bank account, it doesn't belong to you; don't spend it.

As an employer, you must contribute to the Social Security fund for your employees. Your contribution will be equal to the amount withheld from the employee. This is just another hidden expense to having employees. Your portion of the funding will do you no good, it is only to benefit the employee.

Employer ID Number

When you establish your business, you should apply to the Internal Revenue Service for an employer ID number. Some small business owners use their Social Security numbers as an employer ID number, but it is better to receive a formal ID number from the IRS. You will use this identification number when making payroll-tax deposits.

PRO POINTER

Remember the money you withhold from your employee's paycheck is money you use for your payroll-tax deposits. The requirements for when these deposits are made fluctuate from business to business. Consult your CPA for precise instructions on making your payroll-tax deposits.

Payroll-Tax Deposits

Payroll-tax deposits are required by companies with employees. The IRS will provide you with a book of deposit coupons for making these deposits. The deposits can be made at the bank where you do business with.

As a business owner, you can be held personally responsible for unpaid payroll taxes. Even if you sell or close your business with payroll taxes left unpaid, the tax authorities can come after your personal assets to settle the debt. Don't play around with the money owed on payroll taxes.

Federal Unemployment Tax

The Federal Unemployment Tax is also known as FUTA. Your requirements for making FUTA deposits will depend on the gross amount of wages paid in a given period of time. Talk to your accountant for full details on how FUTA will affect your business.

Self-Employment Tax

As your own boss, you may be surprised when forced to pay self-employment taxes. As a self-employed individual, your Social Security tax rate will be double since you will be paying both employee and employer contributions.

State Taxes

State taxes are another consideration for your business. Different states have various rules on their tax requirements. To be safe, check with your CPA or local tax authority to establish your requirements under local tax laws.

Labor Laws

The labor laws control such areas as minimum-wage payments, overtime wages, child labor, and similar requirements. As an employer, you must adhere to the rulings set forth in these laws. Two important pieces of legislation enacted by the federal government impact on your hiring and firing policies. One is the American Disabilities Act (ADA) and the other is the Equal Opportunity Employment Act (EEO). The government will be happy to provide you with information on your responsibilities to these laws. A phone call to the Department of Labor will get details mailed to you.

OSHA

The Occupational Safety and Health Act (OSHA) controls safety in the workplace. Your business may be affected in many ways by OSHA. To learn the requirements of

OSHA, contact the Department of Labor. Most states, nowadays, have their own OSHA departments, so you may be subjected to inspections by either agency.

Terminating Employees

Terminating employees can become a sticky situation. With so many employee rights that may be violated, you must be careful when you are forced to fire an employee. When you hire an employee, start an employment file on the individual. The file will grow to contain all documentation pertaining to that employee. Each file should contain tax forms, employment application, income records, performance reviews, attendance records, and disciplinary actions and warnings. If and when the time comes when you must dismiss an employee, these records of employment history may come in handy.

Before you arbitrarily decide to fire an employee, consider the costs you will incur to replace them. Is it really necessary to fire the individual? If the circumstances demand termination, do so with care. Consult with your attorney in advance and advise them of the circumstances surrounding your decision to let that employee go. If the firing is based on a real or perceived violation of a company policy, make sure you have all of the facts before you make a decision. The classic case involves the firing of an employee who appeared drunk on the job. After being fired, the employee's lawyer presented evidence to the employer that this appearance of intoxication was actually a reaction to a prescription drug. The employer had to pay dearly for his rash decision. Wrongful dismissal of an employee can lead to lawsuits–so check your facts.

PRO POINTER

Since you never know when termination will be your only option in dealing with a troublesome employee, you should assume that all employees can be candidates for termination and therefore you need to create and maintain a paper-trail on each employee.

How to Keep Good Employees

After you have hired your people, you will need to learn how to keep good employees. If employees are worth having, someone else will want them and you will always have to consider the risk that may try to persuade your best employees to leave your company for theirs. You must also be concerned about good employees going into business for themselves and into competition against you. You can't really blame the employees for wanting their own business. After all, you wanted your own business. What you have to do is make the working conditions so good that the employees won't want to leave.

Many factors will influence employees to remain with your company. Some of these factors are:

- Health insurance
- Paid vacations
- Company vehicles
- Good working conditions
- Competent co-workers
- Pride in the company

If you establish a good environment for your employees, they will have no reason to leave your employ. Bonus plans and other incentives can remove some of the risk of having the employees going into business for themselves. It will be up to you to communicate with your employees and to create circumstances to keep them happy. If you have valuable employees, they are worth the extra effort.

Controlling Employee Theft

PRO POINTER

If you eliminate temptation, you eliminate most casual theft.

Controlling employee theft is something most business owners prefer not to think about. All business owners would like to believe that their employees are honest and most are, but a few may not be, so you have to protect yourself and your good employees from the few bad ones.

Most employee theft is petty, but that doesn't mean it is not serious. Stealing is stealing, and you can't afford to have these kinds of people on your payroll. How you run your business will have a bearing on how much employee theft you have. If you screen all of your job applicants thoroughly, you can reduce the chances of hiring someone who is dishonest. Tight control on your inventory, making your employees aware of your anti-theft policies, can help you reduce risks further.

Exercising Quality Control Over Employees

Customers these days are educated consumers and they not only shop price but also look for quality in the product they buy. By educating your employees on the need for quality, you can build a better business. Quality control is just what it sounds like—control over quality in employee performance and quality in the product they produce. The qualities you control could be numerous and may include the following:

- Good work habits
- Work quality
- Loyalty

There are, of course, other qualities you may wish to keep in check. You might encourage your employees to continue their education. And maybe you offer to pay a percentage of any employment related education costs. Once you know what you want from your employees, work with them to help them achieve their goals, which should coincide with yours.

Training Employees to Do the Job

Training employees to do the job used to be standard procedure, but not today. Today, most employers are looking for experienced people that can step into a position and be productive right away. The days of training apprentices are all but gone.

PRO POINTER

To train employees, you must look at the training as an investment.

Why has this shift in the workplace occurred? One reason is money; it costs money to train employees. Even if you do the training yourself, it costs money. The time that you spend away from your routine duties is lost income. This is another reason why employers are reluctant to spend the time training new employees.

In the old days, employees stayed with their employers for a long time. If an employer trained someone, the owner could be reasonably confident the investment would pay off. Today, employees change jobs with the blink of an eye and therefore employers are reluctant to train employees that may leave soon as they have been trained. The traditional values that once existed have been eroded with the increased demand for the mighty dollar. With everyone, both employee and employer in a hurry to grab the brass ring, no one has time to invest in building a stable business or career. Everyone seems to take the shortest path they can find to potential riches.

Should you hire experienced help or train new people to do the job your way? I guess it makes more sense to hire people that can jump out there and start making money for you. But, if you do train employees to produce the type of work you require, you may be happier. There is a certain satisfaction gained from watching a rookie mature into a journeyman. The choices are yours, but remember, some trainees may jump ship.

PRO POINTER

The business world is changing. Since the old school managers are not passing their knowledge down to apprentices, the pool of qualified trades people is shrinking. If this pattern continues, the skilled craftsman of the past will be only a memory.

Training Employees to Deal with Customers

I believe training employees to deal with customers is the responsibility of every business owner. Every business is different, and even experienced mechanics often have to be taught to treat customers the way you want them treated.

It is a good idea to develop a policy manual dealing with building customer relations. Issue this manual to each employee, review and explain the contents, and ask them to absorb what is has to say. If necessary, test the employees' knowledge of the manual. Before you put people in touch with your customers, make sure they will behave in a suitable manner.

PRO POINTER

You may choose not to provide on-the-job training for work skills, but don't forego training your employees to deal with customers. Your customers are your business. If you alienate them, you lose your business. Employees are representatives of your company. If they act improperly around customers, it will be a reflection on your business.

Establishing the Cost of Each Employee to Your Company

Establishing the cost of each employee to your company is a necessary part of setting your pricing structure. There are many hidden costs involved with employees, and each employee may have a different set of circumstances. Before you set your prices, know how much each employee actually costs you, in terms of dollars per hour.

What costs should you look for? The most obvious cost is the hourly wage, but there are other factors. Some employees will receive more benefits than others. One employee with a two-week paid vacation will have more unproductive time, on an annual basis, than other employee who only gets one week of vacation pay. They will be getting paid for 52 weeks, but will only have been productive for 50 weeks.

As employees achieve seniority, they normally gain additional benefits. You must consider all of these costs when determining the overall cost of an employee. Whether it is health insurance, dental insurance, or paid vacation, you must factor the cost into your projections.

Bonuses are employee benefit that affects cost. If you give each employee a bonus during the holiday season, you must include this money in the cost of the employee. Carefully review all of your overhead expenses to sort out those that can be attributed to the hourly cost of your employees. Once you have all the figures, compute the hourly rate for each employee.

PRO POINTER

When you bid a job, include the hourly rate for your most expensive employees. Then, if you assign a lower cost employee to the job, you'll make more money. But, if your less expensive labor is not available, you will not lose money.

Dealing with Production Down-Time When Paying Employees

Dealing with production down-time when paying employees can be frustrating and costly. You have already seen how you can lose money if your crews stop work because they ran out of materials, But that is not the only way you can lose money to down-time. There are situations that you can't control, but the impact can be lessened with the right management decisions.

Bad Weather

Bad weather can often shut a contractor down. While you can't control the weather, you can plan for its effects on your business. If you have a job that involves some inside work, try to save this work for days when the weather won't allow your normal outside operations. When you prepare your estimate, don't assume that you will have all sunny days. Depending upon the time of year, you need to think about adding a small contingency to compensate for time lost to bad weather.

If you have a job that requires you to work through some winter months (if you work in a geographic area where winter means freezing conditions), you need to include some money in your estimate to cover your work and possibly add some heat to the work area. These costs, referred to as "winter conditions", can mount up and if you are required to work continuously through cold snaps, make sure you add some costs for polyethylene and some heaters. Even rain can shut you down so look for ways to keep production up during any weather conditions.

When it is impossible to prevent job shut-downs due to bad weather, use your best judgment in what to do with your employees. Most employers will send the employees home, without pay. On the surface this saves money, but it may cause you to lose your employees. Good employees are hard to find, and a turnover in employees is expensive. You may be money ahead to create some busy work for the crews, even if it is not cost-effective. These tasks may not warrant the use of such highly-paid personnel. But since they must be done, and if it keeps your employees working, you need to establish these types of contingency plans.

PRO POINTER

Some ideas to keep employees busy involve counting or straightening out inventory, taking trucks in for service, or performing maintenance on equipment.

Late Deliveries

Late deliveries can bring your crew's work to a halt. If you are a good manager, you won't let this happen often. There will be times when a delivery will be late, and you must find work for your crews. Be prepared for these times with some back-up plans. If you send the crews home, they may not be happy. On the other hand, maybe they would enjoy having the day off, even if they aren't getting paid. Give them the option of taking the day off or doing fill-in work.

Building Code Violations

Rejected inspections or reported code violations can shut a job down. These things occur because someone is not familiar with the building code, or, out of laziness, does not do the job they are trained to do. If you or your field supervisors are supervising the work, there should be no need for failing an inspection. If the violation was caused by one of your subcontractors, they should be told that they will pay for the cost of lost productivity if any more such violations occur. If you start to have recurrent problems of this nature, you need tighter control on your field supervision.

Disabled Vehicles

Every contractor is going to have problems with disabled vehicles from time to time. The most you can do to prevent these problems is regular maintenance. When a truck breaks down and is going to be out of service for an extended time, try to double your crews up. There isn't much else you can do.

Slow Times or a Lull In Work

Sooner or later, slow times will cause you downtime with your crews. There are certain times of the year when these slow periods are likely to occur. They include holidays, summer vacation seasons, tax seasons, and school start-up seasons.

Proper preparation can help to overcome these slow times. Assign work in anticipation of these slow times. Be aggressive in advertising and offering discounts, if necessary, to keep your people busy.

PRO POINTER

Avoid laying your people off. Once they are gone, you may not get them back.

Reducing Employee Call-Backs and Warranty Work

By reducing employee call-backs and warranty work, you can increase your profits. Customers will not pay you to do the same work twice, but you will have to pay your employees for their time. This can get expensive. If you have sloppy workers that frequently cause call-backs, you must take action.

Everyone is going to make mistakes, but professionals shouldn't make many. Call-backs are generally the result of negligence; either the mechanic did the job too quickly, too poorly, or didn't check the work before leaving the job. You can and must control this type of behavior.

Call-backs and warranty work hurt your business in two ways. The first one is a financial one; you lose money on this type of work. The second problem is the confidence your customers lose in the quality of your work. You cannot afford either of these results. How can you stop call-backs and warranty work? Let's look at some of the ways that have worked for others and that may work for you.

Call-Back Boards

Call-back boards can reduce your call-backs if you have multiple employees. Install a call-back board in a part of your office that all employees can see. When a mechanic has a call-back, the mechanic's name is put on the board. The board is cleared each month, but people with call-backs must see their name on the board for up to a full month.

PRO POINTER

Generally, there is a certain competitiveness among tradespeople. If a mechanic's name is on the call-back board, he will probably be embarrassed. This simple tactic can have a profound effect on your call-back ratio.

Employee Participation

Employee participation in the financial losses of call-backs is another option. However, your employees must agree to this plan without being pressured. For your protection, have all employees agree to the policy in writing. Under the employee-participation program, employees agree to handle their call-backs on their own time. You pay for materials and the employees absorb the cost of the labor.

As a variation of this program, you can agree to pay the employee for the first two call-backs in a given month, with the employee taking any additional call-backs without pay. But, before you implement either of these programs, confirm their legality in your area and have employees agree to your employment terms in writing.

Bonus Incentives

Bonus incentives are another way to curtail call-backs. Call-backs are expensive and detrimental to your business image. If you don't like the idea of giving employees bonuses for doing a job the way they should in the first place, hedge your bets. Determine what the maximum annual bonus for any employee will be, and adjust your starting wages to build in a buffer for the bonuses. The employee will feel rewarded with the bonus and you won't be paying extra for services you expect to get out of a fair day's work.

PRO POINTER

If you can eliminate warranty work by offering bonuses, do it. You won't lose anymore money and you won't lose any credibility with your customers.

Office Employees

Office employees are a little easier to manage than field employees. Office people are relatively easy to find and supervise. If you are an office-based owner, your office employees will stay busy. They know you are watching their performance. However, you must not be overly demanding and abuse the power your presence presents.

Office employees can become uncomfortable if the boss is always looking over their shoulder. Production will drop off or mistakes will multiply. You should hire the best help you can find and then let them do their jobs. It is O.K. to inspect their work from time to time, or offer suggestions to improve their performance.

Although you need to develop a good relationship with your office employees, you should always maintain the separation between boss and office employee. Too much conversation lowers productivity. That doesn't mean you can't discuss the results of the ball game on Monday morning, but make it short and don't encourage lengthy conversations about personal matters. You must manage by example. If you are productive during the day, your employees will follow suit. If you read the paper and drink coffee for an hour or two in the morning, your office people will surely note that and may feel that they too can relax and not jump right into the day's work

When you hire office employees, don't neglect their needs and desires. If you have a good employee that wants a new chair, buy a new chair. When your workers want a coffee maker, buy a coffee maker. If the requests of your help are reasonable, attend to them. Happy employees are more productive, not to mention being more pleasant to be around.

PRO POINTER

Avoid criticizing one employee in front of another. The rule to remember is "Praise in Public- Criticize in Private!"

Field Employees

Field employees present different management challenges than office employees do. These employees are mobile and can be difficult to keep up with. Since the employees are out of the office, their actions are more difficult to monitor. You know every trip your secretary makes to the snack area, but you will be hard pressed to keep up with how many times your field crews take a break.

By monitoring job production, you can keep tabs on your crews. If the work is getting done on time, does it matter if the crew took three breaks, instead of two? If you have good employees that are turning out strong production, leave them alone.

PRO POINTER

Too much employer presence is not good. You are a boss, not a babysitter. When you hire professionals, you should expect them to be competent workers. If you make the decision to hire people, you are going to have to trust them, at least to some extent.

If you are concerned about your field crews, talk to your customers. Customers are generally very aware of how crews are acting. Make some unannounced visits to the job sites. Don't let the crews get too comfortable, but don't crowd them either.

Employee Motivation Tactics

You can use employee motivation tactics to increase the profits of your business. There are many ways to shape employees into a mold that suits you. There are books written for the express purpose of showing employers how to motivate their employees. A creative employer can always find ways to influence employees to do better. Let me give you just a few suggestions that might work with your company.

Awards

Awards are welcomed by everyone. You can issue award certificates for everything from perfect attendance to outstanding achievements and they can make a world of difference in the ways employees act. An employee that knows they will get a certificate for coming to work every day will think twice before calling in sick. While the award may not have a financial value, it becomes a goal. Employees that are working towards a goal will work better.

Money

Money is a great motivator. Since most people work for money, they may work a little harder for extra pay. Any type of bonus program can be beneficial to the production rate of your employees.

A Day Off with Pay

Sometimes a day off with pay is worth much more to an employee than the value of the wages. This special treat could become a coveted goal. One idea would be to hold a contest where the most productive employee of the month gets a day off with pay. Sure, you'll lose the cost of a day's pay, but how much will you gain from all of your employees during the competition?

Performance Ratings

Performance ratings can be compared to awards. If employees know they will be rated on their performance, they may work harder. These ratings should be put in writing and kept in the employee files.

Titles

Wise business owners know that a lot of people would rather have a fancy title than extra money. In fact, some companies award a title, and the prestige that goes with it, instead of a wage or salary increase. Even if your company is small, you can hand out some impressive titles. For example, instead of calling your field supervisor a foreman, call him a field coordinator. Instead of having a secretary, have an office manager. When you have someone that enters data in a computer all day, change the title from data entry clerk to computer operations executive or Information Technology Associate. Titles make employees feel better about themselves, and they don't cost you anything.

If you decide to hire employees, be prepared for some rocky spots in the road. Putting people on your payroll can be very beneficial, but it is not a job that can be done without some sacrifices. Take your time and plan carefully. If you are careful and meticulous, you can avoid most of the problems that so many people have with employees.

CHAPTER THIRTEEN

15 Mistakes That Can Kill Your Business and How to Avoid Them

Over the years, I've done a lot of remodeling and I've come to know more remodelers than I can count. My work as a consultant to other contractors has exposed me to a host of problems. I've suffered with a multitude of difficulties personally, and my work as a consultant has shown me mistakes that I haven't yet made. The combination of my experience as both a remodeler, builder, plumber, and consultant has given me a great deal of respect for the building and remodeling business. It doesn't take but one slip-up to put a contractor out of business. Someone once said that smart people learn from experience, but wise people learn from the experience of others–so hopefully you will learn from other's mistakes.

Contractors just starting out in business will probably make mistakes that are likely to be financially fatal. But experienced remodelers often fall into traps of their own, and I'm not sure that the rookies are at the highest risk. When people start doing something new, they are often careful of every step they take. Once they feel like they are in control, they tend to lighten up on their caution. This is usually when trouble strikes. So even if you've been remodeling houses for 15 years, you could mess up.

Everyone makes mistakes. My mother used to tell me that the first time I did something wrong was experience and that the second time I did the same thing wrong was a mistake. This makes sense but unfortunately business owners sometimes only get one chance to do their job right. This can be especially true for contractors where large sums of money are at risk.

The amount of money at stake when houses are being built and remodeled can be overwhelming and more than the average person could reasonably be expected to pay back if something disastrous happened. I remember a time when I was leveraged out for over four million dollars in construction loans. If something had happened to me at that time of my life, there would have been no way that I could have ever paid the money back by working a normal job.

PRO POINTER

You have taken a major step in the right direction by reading this book. By learning from my experience and mistakes, you will be better prepared to avoid your own. Seeing the danger signs in time to react before you are in too deep is paramount to your success as a builder. Hopefully, you will gain enough knowledge from this book and other sources to make your way to the top of your profession.

There are many mistakes made by contractors and most of the problems can be avoided, but what do you look for? Experience is often the only protection we have against mistakes. How do you survive in a tough business long enough to get the experience you need? It is a difficult situation.

I've thought over all the mistakes that I can remember either making or observed being made. My list is a long one. Many of the problems, however, are not so deadly as to sink your business. So I share with you 15 of the all-time mistakes that I'm aware of. If you study the topics we are about to discuss, you will see how you might avoid mistakes that remodelers frequently make.

It Takes More Money Than You Think

It takes more money to become a full-time contractor than you might think. This is one of the first mistakes remodelers often make. If you sit down and run numbers on what your startup costs will be, the numbers will probably look manageable. The hard-money expense of opening a building business isn't very high. It's the hidden expenses that will put you out of business before you get started.

A lot of people either don't know about or don't think about many of the expenses they will incur when going into business for themselves. For example, you will still have all your personal bills to pay after you quit your job and while you are waiting to draw your first income from your business. That steady paycheck won't be there for you each week.

Income is not the only issue. If you presently have health-insurance benefits supplied by your employer, you might not think to factor the cost of insurance into your new business overhead projections. Forgetting this fact could put you at the unhappy end of a surprise when you realize that hundreds of extra dollars will be needed each month for business insurance premiums.

PRO POINTER

If you don't have enough money saved to support yourself for several months, your time as a professional remodeler could be very short.

Failing to prepare properly when going into business is one of the major reasons why so many businesses never see their one-year anniversary. Take the time to plan out all aspects of your business before you cut your ties with an employer.

Avoiding Heavy Overhead Expenses

Avoiding heavy overhead expenses is something that every business owner should strive to do. Most contractors know that high overhead expenses can be dangerous, but many of them dive right in anyway. If you rush out and rent a lavish office, buy a new truck, lease a heavy-duty copy machine, and spend your money and use up your available credit on things that are not truly needed, you may find yourself in a cash bind. Terminating long leases and installment debt can be very difficult and expensive.

PRO POINTER

Not being able to lower or eliminate your overhead can put you out of business.

It's easy to get caught up in the excitement of owning your own business. Part of the fun is getting to buy things that you want. But you must be reasonable about your purchases and expense commitments. A full-time assistant is something that you probably don't need right away. There are dozens, if not hundreds, of ways to waste your money before you have even earned it. Move cautiously, and don't put yourself in a financial box that you can't get out of.

Too Cautious

There is such a thing as being too cautious. If you tend to be a money miser, you might find that you're not cut out to be an entrepreneur. To make a new business work, you have to take risks and spend money. I know I just finished telling you to be cautious, but don't let fear grasp and immobilize you.

I've seen contractors start a business and refuse to pay one penny for advertising. How they expect to get business if no one knows that they are in business is beyond me. My personal experience and mistakes have made it abundantly clear to me that a person can be at risk by trying to save money without thinking of the consequences. I once terminated a big, expensive ad I had in the phone book to save money. Well, I didn't have to pay the ad expense any longer, but my business dropped off dramatically. I lost much more money in reduced business than I saved. Balance your spending to achieve the best results.

Select Your Subcontractors Carefully

Even though your subs are independent contractors, they have a lot to do with your business image. Sloppy workmanship by your subcontractors will reflect on you. Remember, you are the general contractor and responsible for everything that happens on your job site. If your subs are rude to your customers, you'll take the heat. Many builders are attached to subcontractors who offer the lowest price. This can be a major mistake. Don't assume that price is the only factor when selecting a subcontractor.

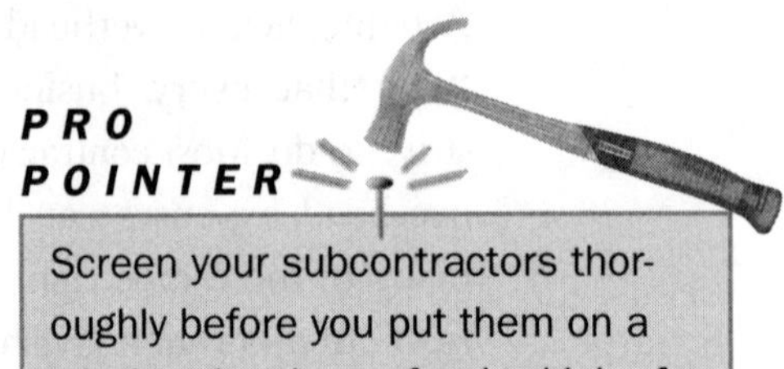

Screen your subcontractors thoroughly before you put them on a job. It only takes a few bad jobs for word to get around that you are a builder to avoid.

Set Up a Line of Credit

Set up a line of credit with your banker before you need it. People often wait until they need money to apply for a loan. This is usually the worst time to ask for money from a bank. My experience with bankers has shown that they are much more likely to approve loans when you don't need them rather than when you do need money.

Most builders run into cash-flow crunches from time to time. Having a line of credit established to tide you over the rough spots until your next draw or payment comes in can make a lot of difference to both your building operation and your credit

CONTRACTOR QUESTIONNAIRE

PLEASE ANSWER ALL THE FOLLOWING QUESTIONS, AND EXPLAIN ANY "NO" ANSWERS.

Company name ________________________________
Physical company address ________________________________
Company mailing address ________________________________
Company phone number ________________________________
After hours phone number ________________________________
Company President/Owner ________________________________
President/Owner address ________________________________
President/Owner phone number ________________________________
How long has company been in business? ________________________________
Name of insurance company ________________________________
Insurance company phone number ________________________________
Does company have liability insurance? ________________________________
Amount of liability insurance coverage ________________________________
Does company have Workman's Comp. insurance? ________________________________
Type of work company is licensed to do ________________________________
List Business or other license numbers ________________________________
Where are licenses held? ________________________________
If applicable, are all workman licensed? ________________________________
Are there any lawsuits pending against the company? ________________________________
Has the company ever been sued? ________________________________
Does the company use subcontractors? ________________________________
Is the company bonded? ________________________________
Who is the company bonded with? ________________________________
Has the company ever had complaints filed against it? ________________________________
Are there any judgments against the company? ________________________________
Please list 3 references of work similar to ours:
#1 ________________________________
#2 ________________________________
#3 ________________________________
Please list 3 credit references:
#1 ________________________________
#2 ________________________________
#3 ________________________________
Please list 3 trade references:
#1 ________________________________
#2 ________________________________
#3 ________________________________
Please note any information you feel will influence our decision:

____

ALL OF THE ABOVE INFORMATION IS TRUE AND ACCURATE AS OF THIS DATE.

DATE:______________ COMPANY NAME: ________________________________
BY:________________________________ TITLE: ______________________

FIGURE 13.1 Example of a questionnaire for a contractor.

report. If you delay paying your suppliers for a couple of months until money comes in, the suppliers may not go out of their way to do business with you in the future. Sometimes jobs are not paid for on schedule, and a remodeler's money can be held up for several weeks or even months.

PRO POINTER

Unless you have sufficient funds in a reserve account, a line of credit in anticipation of problems will save the day.

Get It in Writing

When you are in business, it's always best to get all the details in writing. For a contractor, it's especially important to have contracts with both customers and subcontractors. As a remodeler, you will deal with lots of contracts—why do you think they call us contractors? You should always insist on a letter of commitment that shows a loan has been approved for your customer before you start construction. Some remodelers start work on homes as soon as a customer signs a contract. You can do this, but it's very risky. If your customer is denied a loan, you may never get paid for any of the work or material you have supplied. I hate to say it, but don't take your customer's word for the fact that a loan has been approved. I've never known a lender who didn't issue a letter of commitment once a loan was approved.

PRO POINTER

You should have a copy of the loan commitment letter in your job file. Even if you are working only as a construction manager, you should require your customer to sign a letter of engagement.

Stay Away from Time-and-Material Prices

Stay away from time-and-material (T-&-M) work when dealing with your subcontractors. It is like giving them an open checkbook. It's possible to save money with T-&-M deals, but you can also lose a lot of money. The lure of saving some money is a strong one, but be aware that your attempt to save could backfire and cost you plenty.

I built a house for a customer a couple of years ago and used a site contractor on a T-&-M basis. The contractor was recommended to me by one of my carpenters. After meeting with the contractor, I found that he was just going into business for himself and wasn't sure how to give me a firm price. As experienced as I was, I agreed to a T-&-M basis. I did this to save money and to help the guy out. The deal backfired on me. I'd gotten quotes from other contractors, and the T-&-M price wound up being much higher than any of the quotes. Even though this was a new house, the same type of problem could occur on a remodeling job.

PRO POINTER

Lump-sum pricing is sometimes more expensive, but it's always a safe bet.

Check Zoning Regulations

Before you chop down that first tree or dig your first hole, check the local zoning regulations. This is not a big deal for interior remodeling, but if you are building a garage, a deck, a room addition, or some other types of structure, zoning laws could cause you problems. I know many occasions when seasoned remodelers and builders have gotten into big trouble with zoning problems. There was a commercial building erected in Virginia over a setback line. It had to be moved. A local builder here in Maine recently built a house where part of the foundation was on someone else's land. A carpenter who previously worked for me went into business for himself and placed a well on a building lot that made it impossible to install a septic system without getting an easement for the adjoining landowner. I could go on and on with these types of stories. Granted, these were new construction, but the same type of issue could come up with an addition.

PRO POINTER

Zoning regulations are easy to check into. If you go to your local code- enforcement or zoning office, you can find out exactly what you need to know about setbacks, easements, and other restrictions. If you build a house in a restricted or prohibited space, you're likely to have a major lawsuit filed against you. This can be avoided with some simple investigative work. Something as simple as changing the color of a home's exterior paint could trigger a problem.

Covenants and Restrictions

Many subdivisions have covenants and restrictions that protect the integrity of the development. Customers who buy a home and want to customize the home may not be aware that these restrictions exist. Some builders buy lots without ever asking what restrictions apply. Some examples of common restrictions include a minimum amount of square footage, a prohibition on the use of some types of siding and roofing, and even requirements for the style or type of house that can be built. These rules and laws can affect you, as a remodeler. Don't work on a house and then find out that the siding you used has to be torn off and replaced. Even more importantly, don't build a two-story garage and then discover that the subdivision only allows one-story garages. This can happen. You can check for covenants and restrictions yourself by reading a copy of the deed to a piece of property

Insurance

Some states require contractors to carry a minimum amount of insurance, and others don't. Whether your local laws require insurance or not, you should get it. At the very least, you need liability insurance. There are many other types of coverage that may be needed, such as worker's compensation insurance. Talk to your insurance agent for advice on exactly what types of coverage you should have.

PRO POINTER

If you work without the proper property and general liability insurance, like a few contractors I know, you are sitting on a time bomb. One lawsuit is all it takes to ruin your life. You could lose your business, your home, and your future. Insurance premiums may seem like an unnecessary expense to pay, but if you ever need the coverage, you'll be thankful that you have it.

Inaccurate Quotes

Inaccurate quotes can have you working for nothing. Most contractors commit to jobs for a lump sum or flat fee. If the price you quoted is too low, too bad for you. While you may not bid a job so low that you actually lose money, it's very possible that you will give away some percentage of your profit due to mistakes in your pricing.

Contracting is a business that often runs in cycles. You might go for months with no work in sight and suddenly get flooded with requests for quotes. This is a dangerous situation. You've been sitting around with nothing to do and wondering how you will ever pay the bills. When you finally get a chance to bid a job, you might tend to submit a low bid to make sure you get it. This may be bad enough, but it could get worse. When bid requests are piling up on your desk, you might rush through them, hoping to make up for lost time. You can make mental errors under these conditions and lose a lot of money.

If you make mistakes in your take-offs or estimates, you may not know it until a job is nearing completion. By then, it's usually too late to do anything other than reconcile yourself to losing money. Many remodelers, myself included, have lost enough money by making mistakes when quoting jobs, and I've lost lots of sleep over the experience. Take your time and make sure your estimates are right. Double-check everything you do. Once you commit to a price in a contract, you will be stuck with it.

PRO POINTER

If you have someone who can review your numbers, such as one of your field supervisors, do it. Fresh eyes often catch omissions.

Inspect Your Jobs Frequently

Inspect your jobs frequently. Some contractors are reluctant to go out into the field on a regular basis to check on schedules and quality of work. They're either too busy, don't want to run into the customers, or are just too lazy. You need to visit your job sites often. I try to get to every job at least once a day. When I was doing volume building, I had two full-time supervisors checking every job twice a day.

PRO POINTER

Failure to inspect jobs results in poor quality control. When this happens, a remodeler's reputation can be tarnished.

A lot can happen in just one day. If you skip an inspection, your customer may know more about the condition of a house than you do. It's very embarrassing to have a customer come to your office making a complaint about a condition that you are not aware of. As the general contractor, you should stay on top of your jobs as best you can. If you get a bad reputation in the business, you might as well look for some other type of work.

Customer Relations

Customer relations are a critical part of a contracting business. A substantial remodeling job is a very emotional experience for the homeowner and sometimes he or she will become irrational. You have to be willing to talk with your customers, and you may have to settle them down from time to time. If you don't pay attention to your customers, you won't get the coveted referral business that so many contractors live off. It's essential that you keep your customers happy.

Change Orders

If your customers ask for added services or changes in contract obligations, use change orders to document the requested change or additional service you are asked to provide. Otherwise, you could be left holding the bag on some expensive items. A customer who upgrades from a regular bathtub to a whirlpool tub could be increasing the cost of a home by thousands of dollars. If you don't have authorization in writing to make this change, you might never get paid for the increased cost of your plumbing contract.

I can't remember many jobs that I've done where a customer didn't request some type of change or addition to the original contract. In my early years, I would accept a verbal authorization for changes, but not any more. Too many times I did the work as requested and never got paid. Collecting your money for extras can be difficult enough when you have written documentation of authorization, and it can be almost impossible without some type of proof. A lot of remodelers lose a lot of money by not putting all their agreements in writing. Don't allow yourself to become another statistic in the book of contractors who lost money.

Never Get Too Comfortable

Never get too comfortable with your business. Remodelers who think they know it all often find out quickly that they don't. If you slack up, the competition will take advantage of your reduced effort. To enjoy continued success year after year, you can never stop improving your business. Some businesses, such real-estate brokerages, must maintain continuing-education requirements. You may not fall under such regulations, but you owe it to yourself to always improve your knowledge and understanding of what it is that you do for a living. If you stop learning, you will likely see a decline in your profits. At the very least, you probably won't see any increase in your business. Regardless of how long you work as a builder, there are always new things to learn.

CHAPTER FOURTEEN

Bidding Jobs at Profitable Prices

Building a business clientele is one of your most important jobs as a business owner. Learning how to prepare and submit winning bids is one of the most effective ways to build your business clientele. How many times have you bid a job and never heard back from the potential customer? Many contractors never figure out how to win bids successfully. This chapter is going to show you how to submit bids that build your business.

Word-of-Mouth Referrals

Word-of-mouth referrals are the best way to get new business. Of course, you will need some business before you can benefit from word-of-mouth referrals. But every time you get a job, you need to think about using that customer as a referral for future contracts. A satisfied customer is your best route to attracting new clients.

Advertising is expensive. For every job you get from advertising, you are losing a percentage of your profit to the cost of advertising. If you can develop new work from existing customers, you eliminate the cost of advertising.

If you do quality work and live up to your commitments to your customers, referrals will be easy to come by, but you may have to ask for them. People will sometimes give your name and number to friends, and they occasionally write nice letters. However, to make the most of word-of-mouth referrals, you have to learn to ask for what you want. Let's see what it takes to get the most mileage out of your existing customers.

PRO POINTER

Getting referrals from existing customers is not only the most effective way to generate new business, it's the least expensive.

Laying the groundwork is an important step in getting a strong portfolio of customer referrals. If you don't make your customers happy, they will talk to their friends, but they won't be saying too many kind words about you or your company. People are quick to talk about their bad experiences, but they may not be so quick to spread good words. To get the message out that you want referrals, you have to work hard at pleasing your customers.

Start with the first contact you have with customers and work to continue that relationship. Call them from time to time after you have completed the work to find out how they are enjoying that new addition or the new bathroom or that new house. Ask about the family, how the kids are doing. This show of interest will not be lost on your customers and when asked for a referral they will respond with a positive one.

I have seen many good jobs go bad in the final days of their completion. One of the biggest mistakes a contractor can make is not responding promptly to warranty calls. All the goodwill you have created throughout the job may go down the drain because you failed to respond promptly to that leaking faucet, that toilet that won't stop running, or that bedroom door that won't close properly. If you are in business for the duration, you want to please your customers, not alienate them, particularly after the job is done. Old customers become repeat customers.

PRO POINTER

If you don't respond quickly and effectively to callbacks, you won't be called when there is new work to be done.

During the job you must cater to the customers by being punctual at all your meetings, having answers to their questions, listening to any problems they may have observed, being respectful, and being professional. Always keep in the back of your mind the thought that you want these customers to be happy with your work because you'd like to use them as a reference.

At the completion of the job you should ask your customers for a reference. Don't expect them to offer to give you one. Often, asking for a letter of reference isn't enough. It helps if you provide a form for the consumer to fill out. People never seem to know what to say in a reference letter. They are much more comfortable filling out forms.

If you design a simple form, almost all satisfied customers will complete and sign it. I'm sure you have seen these quality-control forms in restaurants and with mail-order shipments. You can structure the form in any fashion you like.

Once you have created and printed your forms, use them. When you are completing a job, ask your customer to fill out and sign your reference form. Do it on the spot. Once you are out of the house, getting the form completed and signed will be more difficult.

As you begin building a collection of positive reference forms, don't hesitate to show them to prospective customers. Use an attractive three-ring binder and clear protective pages to store and display your hard-earned references. When you get enough reference letters, you will have strong ammunition to close future deals.

Customer Satisfaction

Customer satisfaction is the key to success. Business builds upon itself when customers are satisfied. Of course, there are some people that you may never be able to satisfy, and these hard-to-please people seem to exist for every business owner, but they are the exception rather than the rule Let's concentrate on pleasing the bulk of your customers.

PRO POINTER

Some customers will require a lot of hand holding, and, on occasion, you may have to spend more time with these clients in order to make them feel comfortable with the decisions they have made or are planning to make.

Customers like to feel comfortable with their contractors. To make customers comfortable, you have to deal with them on a level that they feel at ease with. Remember, they may not be familiar with construction or construction terms so you have to speak to them in terms that they understand. Communication skills are essential in building good relationships. If you and the customers can't communicate easily, you'll have a much more difficult time in your future business dealings.

Reaching Out for a New Customer Base

You can start reaching out for a new customer base through the use of bid sheets, a system of notification of pending jobs in either the public or private sector that are advertised. Bid sheets are a way of notifying contractors of potential projects, and any contractor can bid on them. Listing a project on a bid sheet is generally recognized as a commitment by the owner to proceed with the work if bids come in within their budget, unlike some estimates requested by residential owners that may be requested for information purposes only and do not represent a real job. This type of competitive bid work frequently results in a job with a low profit margin, but if the estimate is correct it can still pay the bills.

Where are Bid Sheets Available?

Bid sheets can be obtained by responding to public notices in newspapers and by subscribing to services that provide bid information. If you watch the classified section of major newspapers, you will see advertisements for jobs going out for bids. You can receive bid packages by responding to these advertisements. Normally, you will get a set of plans, specifications, bid documents, bid instructions, and other needed information. These bid packages can be simple or complicated, depending upon the nature and size of the project.

PRO POINTER

There is a difference between a bid sheet and a bid package. The bid sheet will give a brief description of the work available. A bid package gives complete details of what will be expected from bidders. Most contractors check bid sheets and if they find a job of interest, order a bid package. Bid sheets can be either free if posted in a newspaper or magazine or require a check if provided by a private service.

What is a Bid Sheet?

A bid sheet is a formal request for price quotes. A typical bid sheet contains bidding information similar to this:

> **Requests for Bids:**
> A 3,500-square-foot, two-story new residence is being planned for a 1/2-acre parcel located on the corner of St. Paul and Charles Streets, Baltimore, Maryland. Interested contractors can obtain the plans and specifications by contacting Mr. James Smith, architect, at the office located at 555 New Bern Avenue, Towson, Maryland, tel: 410 854-4444. A refundable check in the amount of $125 is required for the bid package. Bids are due on April 15.2004.

Bid-Reporting Companies

There are several companies that are in the business of providing bid information. Perhaps the best known is the F.W. Dodge Company. Local addresses of these types of companies can be found in most Yellow Page telephone directories or over the Internet. These listings are normally published in a newsletter form. The bid reports are generally delivered to contractors on a weekly basis. Each bid report may contain five jobs or fifty jobs. These publications are an excellent way to get leads on all types of jobs.

All types of projects are included in these bid sheets. They range from small residential jobs to large commercial jobs. The majority of the jobs are commercial. The size of the jobs ranges from a few thousand dollars to millions of dollars.

Public-Agency Bid Sheets

Bid sheets prepared by local, state, or federal government agencies present another opportunity to find work. Like other bid sheets, government sheets give a synopsis of the job description and provide information for obtaining more details. Government jobs can range from replacing a dozen lavatory faucets to building a commissary. Building new base housing units for the armed forces could result in many months of work for a homebuilder. In the same vein, remodeling aging homes can result in substantial work for remodelers.

Government agencies represent a fairly stable type of project. However, there is a lot of paperwork and sometimes payments are slow in coming. There are several things to consider when deciding to bid on a government project: there will be lots of forms to fill out and lots of paperwork, a bond will be required, and there are many laws and ordinances that will have to be followed.

PRO POINTER

If you are not willing to deal with mountains of paperwork, stay away from government bids.

If public funds are used in any of these types of projects, they will fall under the state and federal affirmative action and equal-opportunity laws. If you qualify, provisions in these types of contracts to eliminate discrimination and its effect may be beneficial to you. A typical clause in those bid documents would be similar to this one:

> For the purpose of this contract , a goal of "X" % has been established for socially and economically disadvantaged businesses that are owned and controlled by those individuals who are Black Americans, Hispanic Americans, Asian-Pacific Americans, Subcontinent Americans, Native Americans, or Women, pursuant to (and they name the laws).

A company certified by the state and/or federal government as an MBE (Minority Business Enterprise) or a WBE (Women-Owned Business Enterprise) is encouraged to bid, and a specific number of these firms will be hired for portions of the project. If you feel you can qualify for any of these designations, a little investigation might be beneficial to you if you wish to seek work in the public sector.

Payment, Performance, and Bid Bonds

Payment, performance, and bid bonds are a necessity with many major jobs. If you order a bid sheet or package, you will almost certainly see that a bond is required. Some listings on bid sheets may not require a bond. Requirements for bonds are often tied to the anticipated cost of the job. The bigger the job, the more likely it is that a bond will be required. All public projects will require bonds.

Are You Bondable?

Many of the jobs listed on these bid sheets require contractors to be bonded. Bonds are obtained from bonding and insurance companies. Obtaining "bondability" is not easy. Before you try bidding jobs that require bonding, check to see if you are bondable. The requirements for being bonded vary. Check in your local phone book for an insurance agent that also handles bonds for contractors, and call to inquire about the requirements.

Why Are Bonds Required?

Bonds are required to ensure the successful completion of a project. A payment bond assures the owner of the project that all suppliers, vendors, and subcontractors will be paid at the end of the project. A performance bond assures the owner that the contractor will "perform" in accordance with the terms and conditions of the contract. If the contractor defaults on either of these obligations, the bonding company steps and provides the necessary funds to satisfy these defaults. When the person or firm awarding the contract requires a bond, they know there is a degree of safety.

Bid bonds are slightly different. Bid bonds are generally required in an amount equal to 10 percent of the project's cost. They are submitted with the contractor's bid. If the contractor is notified that he is the successful bidder, but declines to accept a contract, the owner will make the award to the second bidder. The proceeds from the bond will be used to make up the difference if any between the low bidder's price and the second bidder's price. The contractor who declined to accept the bid will lose his bond and forfeit the money it cost.

PRO POINTER

If you are required to submit a bid bond, make sure that you are willing to accept a contract if offered or else you'll pay the penalty.

It is very difficult for some new businesses to obtain a bond. If the new company doesn't have strong assets or a good track record, getting a bond is tough. But when you put up a bond, the value of the bond is at risk. If you default on your contract, you lose your bond to the person who contracted you for the job. Since many people use the equity in their home as collateral for a bond, they could lose their house. Bonds are serious business. If you can get a bond, you have an advantage in the business world. Talk with local companies that issue bonds to see if you can qualify for bonding.

Big Jobs—Big Risks?

Do big jobs carry big risks? You bet they do. There are risks in all jobs, but big jobs do carry big risks. Should you shy away from big jobs? Maybe, but if you go into the deal with the right knowledge and paperwork, you should survive and possibly prosper.

Cash-Flow Problems

Cash-flow problems are looming when contractors take on big jobs. Unlike small residential jobs, big jobs don't generally allow for contractors to receive cash deposits. If you tackle these jobs, you will have to work with your own money and use your credit lines. When involved in a big project, each request for payment will be large and some will be huge. If payments are late and you have to pay out thousands of dollars to suppliers and subcontractors, they will only wait so long for their money before they pull off the job. You'll need the financial resources to pay these people if your customer is late with her payments. I once knew a very successful contractor who was given the opportunity to work on a multimillion-dollar project. He turned it down. When I asked him why, he said, "You know if these guys are late on one payment, it will drain my resources and hurt my other jobs–I think I'll pass." Wise decision. It's wonderful to think of signing a million-dollar job in your first year of business, but that job could put your business into bankruptcy court. Some lenders will allow you to use your contract as security for a loan, but don't bet your business on it.

If you want to take on a big job, get your finances in order first.

Slow Pay

Slow pay is another problem, not necessarily related to big jobs, but you undoubtedly will run into slow payers during your business career. Keeping track of when your payments are due is extremely important when you have tackled a big job, but it can be equally important if you have a number of small jobs where the owners are lax in their payments. New businesses are especially vulnerable in succumbing to slow pay. The check you thought would come last month might not show up for another 60 or 90 days, so stay on top of your accounts receivable and don't be afraid to call that customer who is overdue in paying; he will probably think more of you as a businessperson who pays attention to business.

No Pay

It is not unknown for an owner to suddenly declare bankruptcy. While it is never good, it always seems to happen at the worst possible moment–just when you have submitted that big pay request. You can request a copy of the owner's financial commitment from a bank or other lending institution. If privately financed, it becomes more difficult, but if you notice a trend toward slow payments, you'd better investigate. It may be nothing or it may be the beginning of a bankruptcy proceeding . Some contractors are reluctant to ask an owner for proof of financing, but many owners will appreciate your businesslike approach.

If the subcontractors don't get paid, suppliers don't get paid. The ripple effect continues. Anyone involved with the project is going to lose. Some will lose more than others. Generally, when jobs go bad, the banks or lenders financing the whole job will foreclose on the property. These lenders normally hold a first mortgage on the property.

PRO POINTER

Knowing the financial situation of your customer is a first step in avoiding a "no pay" situation.

If you're working as a subcontractor, filing a mechanic's lien is the best course of action when your customer refuses to pay you. If as a general contractor you haven't been paid for labor or materials, a mechanic's lien can usually be levied against the property where the labor or materials were invested. If you have to file a lien, make sure you do it right. There are rules you must follow in filing and perfecting a lien. You can file your own liens, but I recommend working with an attorney on all legal matters. Here are some important items to remember:

- Check with your local government to find the exact requirements for filing a lien.
- You usually have to file a lien within a certain period of time after you last worked on the project, but be careful about this requirement. Let's say the time required to file the lien is 60 days after you have last worked on the project. Say it is day 59 and you haven't filed, so you go back to the project and clean out the strainer on the kitchen faucet. This won't work. Most laws state that you have to perform "meaningful work" within that time frame, and cleaning a strainer won't qualify
- Since the lien will be filed against the owner of the property, you need to ensure that you have the correct name of the owner and the correct description of the property. This can generally be found in the land records at the Town Clerk's office.

If the lien is not "satisfied" (paid by the owner), the court will allow a judgment to be filed against the property and foreclosure will be the final step. If you get any money, it will likely be a settlement in a reduced amount. You will never quite get

PRO POINTER

The actual preparation and filing of a mechanic's lien is best left to your attorney.

the taste of that bad job out of your mouth, but you should learn from it .Did I let that first past-due payment go by without calling the owner? Were there signs that things were going bad, but I either didn't recognize them or thought that things would turn out O.K? Try to learn from your mistakes.

Completion Dates

When you sign a contract with a specific completion date, you had better be prepared to meet that date. Some contracts have penalty clauses for late delivery. Even if your contract does not penalize you, your reputation is on the line. When you commit to a completion date- make sure you make it.

Contracts for large projects frequently include liquidated damages, referred to as LDs, which are assessed for each day the project finishes beyond that date stated in the contract. The daily amount of the LD is meant to represent the cost to the owner for not having the project completed on time. These liquidated damages can run from $50 per day to $5,000 per day or more. Liquidated damages and the possible loss of your bond can ruin your business.

PRO POINTER

Contractors with limited experience on big and small jobs are often unprepared to project solid completion dates. Don't sign a contract with a completion date you are not sure you can meet.

The Bid Process

When you learn how to beat your competition in the bid process, you are on your way to a successful business. There is no shortage of competition in most fields of contracting. Some contractors can get discouraged. But don't—there are ways to forge ahead of the competition.

The Competitive Nature of the Bid Sheet

Beating the competition when projects are obtained from bid sheets is tough. The low bidder gets the job and with three or four, or sometimes a dozen, bidders competing for the work, someone can make a mistake in his estimate or decide to submit a "lowball' bid to get the job for one reason or another. Being the low bidder can get you the job, but you may wish you had never seen the job.

If you can get bonded, you have an edge. A lot of bidders can't. This fact alone can be enough to cull the competition. When you prepare your bid package for submission, be meticulous. If you want the job, spend sufficient time preparing a professional bid packet. Discuss the job with several qualified subcontractors to get the best price. Shop for materials and equipment, and double-check your estimate to see if you're left something out or included something you shouldn't have. Check for arithmetical errors—did you multiply 10 x 35 and get 3500?

PRO POINTER

Don't bid a job too low. It doesn't do you any good to have work if you're not making money.

Submitting Bids In-Person

If you are dealing with bids delivered to the customer in person, follow the guidelines found throughout this book. The basic keys include: dress appropriately, drive an appropriate vehicle, be professional, be friendly, get the customer's confidence, produce photos of your work, show off your letters of reference, give your bid presentation in person, and follow up on all your bids.

Preparing Accurate Take-Offs

To prepare accurate bids, you must be adept at preparing accurate take-offs. It doesn't matter if you use a computerized estimating program or a pen and paper—you must be precise. If you miss items in the take-off and get the job, you will lose money. If you overestimate the take-off, your price will be too high.

PRO POINTER

An accurate take-off is instrumental in winning a job.

What is a Take-Off?

A take-off is a list of items required to do a job. Take-offs from blueprints or visiting the job site produce a list of everything you will need to do the job. Some estimators are wizards with take-offs, and others have a hard time trying to figure out all the materials and components needed. If you can't discipline yourself to learn how to do an accurate take-off, you will need to either hire someone who can or find an estimating service that will provide you with an accurate estimate for a fee.

Using Take-Off Forms

You can reduce your risk of errors by using a take-off form. If you use a computerized estimating program, the computer files probably already contain forms. You may want to customize the standard computer forms. Whether you are using standard computer forms or making your own forms, you must be sure that they are comprehensive.

Take-off forms should include every item required for the type of job being estimated. It's best if you create forms that list every expense that you might incur on a job. This improves your odds of reducing errors and omissions when figuring prices. In addition, there should be a place on the form that allows you to fill in blank spaces with specialty items.

Materials Take-off List.

Item name or use of piece	*No. of pieces*	*Unit*	*Length in place*	*Size*	*Length*	*No. per length*	*Quantity*
1. Footers	45	Pc	1'5"	2×6	10'	7	7
2. Spreaders	30	Pc	1'4"	2×6	8'	6	5
3. Foundation post	15	Pc	3'0"	6×6	12'	4	4
4. Scabs	20	Pc	1'0"	1×6	8'	8	3
5. Girders	36	Pc	10'0"	2×6	10'	1	36
6. Joists	46	Pc	10'0"	2×6	10'	1	46
7. Joist splices	21	Pc	2'0"	1×6	8'	4	6
8. Block bridging	40	Pc	1'10⅜"	2×6	8'	4	10
9. Closers	12	Pc	10'0"	1×8	10'	1	12
10. Flooring	800	BF	RL	1×6	RL	—	—

FIGURE 14.1 Example of a take-off list.

Job Take-Off Form

Job Name: ______________________

Job Address: ______________________

Item	Quantity	Description
2" pipe	100'	PVC
4" pipe	40'	PVC
4" clean-out w/plug	1	PVC
2" quarter-bend	4	PVC
2" coupling	3	PVC
4" eighth-bend	2	PVC
Glue	1 quart	PVC
Cleaner	1 quart	PVC
Primer	1 quart	PVC

FIGURE 14.2 Example of a take-off form.

Take-off forms alert you to items you might otherwise forget. However, don't just look for items on your form. It is very possible a job might require something that you haven't yet put on the form. Forms help, but there is no substitute for thoroughness.

Keeping Track

Keeping track of what you've already counted is a problem for some contractors. If you are doing a take-off on a large set of plans-say, a shopping mall–it can be tedious work. The last thing you need to have happen is to lose your place or forget what you've already counted. To avoid this problem, mark each item on the plans as you count it. Using various colored pencils or highlighters is an easy way to keep track of what you have or have not counted. You might use a red pencil to check off each door and a blue one for each window. You might use a yellow highlighter on concrete work or a green highlighter on underground storm or sanitary lines.

Build in Contingencies

You should build a contingency in your take-off. If you think you are going to need 100 sheets of plywood, add a little to your count. How much you add will depend on the size of the job you are figuring. A lot of estimators figure a contingency of 3 to 5 percent. Some contractors add 10 percent to their figures.

PRO POINTER

Unless the job is small, I think a 10-percent add-on can cause you to lose the bid.

And don't forget to add something for waste. Rarely will you use every bit of your framing lumber, plywood, or sheetrock. You'll need to buy a certain quantity, but some will get thrown out in the trash. With accurate take-offs, a small percentage for oversights will be sufficient. If you always seem to get on the job and run short of materials, a higher contingency will be needed.

Keep Records

Keep records of your material needs on each job. Don't throw away your take-off sheets. When the job is done, compare the material actually used with what you estimated in your take-off. This will not only help you to see where your money is going. It will make you a better estimator. By tracking your jobs and comparing final counts with original estimates, you can refine your bidding techniques and win more jobs.

Pricing

PRO POINTER

Some newcomers to the contracting business make a serious mistake. When they learn what their competitors are charging, they price their services far below the crowd, hoping to attract some business.

Pricing your services and materials is an essential part of running a profitable business. If your prices are too low, you may be very busy, but your profits will suffer. If your prices are too high, you will be sitting around, staring at the ceiling, and hoping for the phone to ring. Somewhere between too low and too high is the optimum price for your products and services. The trick is finding out where that level is.

You must learn how to make your prices competitive without giving away too much of your profit. What's the right price? Well, you can't pull your prices out of thin air. You must establish your pricing structure with research—lots of research.

COST ESTIMATES FORM

Cost Projections For Bathroom Remodeling

Item/Phase	Labor	Material	Total
Plans			
Specifications			
Permits			
Trash container deposit			
Trash container delivery			
Demolition			
Dump fees			
Rough plumbing			
Rough electrical			
Rough heating/ac			
Subfloor			
Insulation			
Drywall			
Ceramic tile			
Linen closet			
Baseboard trim			
Window trim			
Door trim			
Paint/Wallpaper			
Underlayment			
Finish floor covering			
Linen closet shelves			

(continues)

FIGURE 14.3 Example of a cost estimates form. *(continued on next page)*

Setting extremely low prices can be the same as setting a trap for yourself. You may start a price war with the area's established competitors, driving prices and profit margins even lower, and you may alienate prospective customers. As a new company with low overhead, it's O.K. to price your product lower than the well-established companies, but don't price yourself into a deep hole.

When it comes to determining attractive prices, you must look below the surface. There are many factors that control what you are able to earn. Let's take a look at what is considered a profitable markup.

What Is a Profitable Markup?

What is a profitable markup on materials? This can be a hard question to answer. It is not difficult to project what a reasonable markup is, but defining a profitable markup is not so easy. Some contractors feel a 10-percent markup is adequate. Others try to add 35 percent onto the price of their materials. Which group is right? Well, you can't make that decision with the limited information I have given you. The contractors who charge a 10-percent markup may be doing fine, especially if they deal in big jobs and large quantities of materials. The 35-percent group may be justified in their markup, especially if they are selling small quantities of lower-priced materials.

PRO POINTER

Markups will need to be adjusted to meet the changes in market conditions and individual job requirements.

Markup is a relative concept. Ten percent of $100,000 is much more than 35 percent of $100. You can, however, pick percentage numbers for most of your average sales. If you were in a repair business and were typically selling materials that cost around $20, a 35-percent markup would be fine if the market will bear it. If you are building and selling houses, a 10-percent markup on materials should be sufficient. To some extent, you have to test the market conditions to determine what price consumers are willing to pay for your materials.

PRO POINTER

If you typically install specialty items, you can increase your markup.

If you are selling everyday items that anyone can purchase from a local hardware or building-supply company, you must be careful not to inflate your prices too much. Customers expect you to mark up your materials, but they don't want to be gouged. If you installed light bulbs in that new light fixture and charged twice as much for the bulbs as the customer could have bought them for in the store, that customer is not going to be too happy even though the amount of money involved in the light-bulb transaction is small. The principle of being charged double for a common item still exists.

People will not be as upset when asked to pay a high markup specialty items. A markup of 20 percent will almost always be acceptable on small residential jobs. When you decide to go above the 20-percent point, do so slowly and while testing the response of your customers. Realistically, you will probably find that a 10-percent markup will work best. You might be able to increase it to 15-percent, but that's usually about the limit for big-ticket items.

How Can Your Competitors Offer Such Low Prices?

This is a question almost all contractors ask themselves. There always seems to be some company out there that has a knack for winning bids and beating the competition. You can't help but wonder how that company does it.

Low prices can keep companies busy, but that doesn't mean the low-priced companies are making a profit. Gross sales are important, but net profits are what business is all about. If a company is not making a profit, there is not much sense in operating the business.

Companies that work with low prices fall into several categories. Some companies work on a volume basis. By generating a high volume, the company can make less money on each job and still make a decent profit by doing so many jobs. This type of company is hard to beat because of its high volume.

When I was building in Virginia, I operated on a volume basis. My profit from a house was only about $7,000, but I was building as many as 60 of them a year. If you do the math, you'll see that I was making a good living. The volume principle worked well for me in Virginia, but it won't work for me in Maine. I probably built more houses in a year while working than Virginia than are built by all the builders in my area of Maine in any given year. There simply isn't enough demand for housing in Maine to allow the volume approach to work very well.

Some companies sell at low prices because they are not really aware of their overhead and how much it really costs them to do business. For example, a carpenter who is making $16 an hour at a job might think that going into business independently and charging $25 an hour would be great. The carpenter might even start by charging only $20 an hour. From the carpenter's perspective, he is making at least $4 an hour more than he was at his job. Is he really making $4 an hour more? Yes and no. He is being paid an extra $4 an hour, but he is not going to get to keep much of it.

When this carpenter with his low hourly rate starts up in business, he may take a lot of work away from established contractors. Experienced contractors know they can't make ends meet by charging such low hourly rates. But our carpenter in this example, inexperienced as a business owner, is not aware of all the expenses he is incurring. Once the overhead expenses start eating away at what he thought was a great profit, he may be in for a real shock when he finds out that he is actually working for nothing. This type of businessperson will soon be out of business.

PRO POINTER

A businessperson will soon learn that overhead expenses are a force to be reckoned with. Insurance, advertising, warranty and callbacks, self-employment taxes, and a lot of other expenses not initially considered will erode any profits quickly.

For established contractors competing against newcomers, being able to endure the momentary drop in sales will be enough to weather the storm. In a few months the new business will either be gone or have increased its prices to a more realistic level.

Pricing Services for Success and Longevity

Pricing services and materials correctly is related directly to your success and longevity. It can be very difficult to determine what the right price is for your time and materials. There are books with formulas and theories about how to set your prices, but these guides are not always on the mark. Every location can dictate variations in the prices the public is willing to pay. There are several methods that you can use to find the best fees for your business to charge. Let's look at some of them.

PRO POINTER

The idea behind estimating guides is a good one, but they are only guides. I have read and used many of these pricing books. From my personal experience, the books have not been accurate for the type of work I was doing. There are many times when the books are accurate, but I have never been comfortable depending solely on a general pricing guide.

Pricing Guides

Pricing guides can be a big help to the business owner with little knowledge of how to establish the value of labor and materials. However, these guides can cause you frustration and lost business. Most estimating guides provide a formula for adjusting the recommended prices for various regions. For example, a two-car garage that is worth $7,500 in Maine may be worth $10,000 in Virginia. The formula used to make this type of adjustment is usually a number that is multiplied against the recommended price. By using the multiplication factor, a price can be derived for services and materials in any major city.

Estimating books are very helpful, and while I don't use pricing figures from these books as my only means of setting a price, I do use them to compare my figures and to ensure that I haven't forgotten items or phases of work. Many bookstores carry some form of estimating guides in their inventory, and these types of books can be purchased on the Internet.

Research

PRO POINTER

Real estate appraisers are an excellent source for pricing information.

Research is an effective way to determine your pricing. When you look at historical data, you can find many answers to your questions. You can see how the economy swings up and down between supply and demand, between a buyer's market and a seller's market. You can see how prices have fluctuated up and down over the years, and you may be able to project the curve of the future. Historical data can be found by checking old newspaper ads on file in your local library, researching tax evaluations on homes, and talking with real-estate appraisers. Most appraisers are willing to consult with contractors on an hourly basis.

Let's say that you are a general contractor and that you remodel houses and build garages, additions, decks, and related home improvements. One way for you to establish the value of these services is to consult with a licensed real-estate appraiser. The appraiser can tell you what value the service will have on an official appraisal of the property. This doesn't guarantee that the values given by the appraiser are the best prices for you to use, but they are an excellent reference. By spending a little money to talk with an appraiser, you can save thousands of dollars in lost income. If your business involves providing goods and services for homes and businesses, appraisers are one of your best sources of pricing information.

Assume that the appraiser gives you a value of $2,000 for a 10-foot-by-10-foot deck. You could ask the appraiser to provide a written statement of value. Of course, the value will be general in nature and may not apply to all conditions, but the $2,000 figure will be a solid average. You can use this written statement as a sales tool when you sit down with customers. If you are willing and able to sell the deck for $1,800, you can show the customer that it is lower than the average retail value of the deck.

You can also use the appraiser's figure in conjunction with numbers given in pricing guides and your own estimate to arrive at an average cost. In the case of new houses, the appraised value will typically be the selling price of your house. Unless your buyer is willing to put up cash for any overage, you will be limited to the appraised value.

Combined Methods

Combined methods are best when establishing your pricing principles. Use as much research and as many resources as possible to set your fees. Once you feel comfortable with your rates, test them. Ask customers to tell you their feelings. There is no feedback better than that from the people you serve.

Proper Presentation

Proper presentation is critical for business success. Even if your price is higher than the competition, you can still win the job with an effective presentation. There are many occasions when the low bidder does not get the job. As a contractor, you can set yourself apart from the crowd by using certain presentation methods. There are numerous ways to achieve an edge over the competition. The ways to win the bid battle can include the way you dress, what you drive, your organizational skills, and much more.

When you learn to convince a customer why you are worth more, you will make the most for your time and effort. People are often willing to pay a higher price to get what they want, and it will be your job to convince the customer that you can provide better products and services. Let's look at some methods that have proven effective over the years.

Mailing Follow-Ups

You might be surprised by how many contractors mail estimates to customers and wait to be contacted. Many never hear from the customer again. Mailing bids to potential customers may not be the most effective means. If customers are soliciting bids from your competitors as well, they may only look at the bottom line when making a decision. You want to make customers aware of the added benefits you can supply, so you need a better method of presentation other than a cold mailing.

> **PRO POINTER**
>
> Use the phone to get leads, set appointments, and follow up on estimates, but don't use it to give prices and proposals to possible customers.

If you must mail your proposal, make sure you prepare a professional package. Use printed forms and stationery, not regular paper with your company name rubber-stamped onto it. Use a heavyweight paper and professional colors. Type your estimates and avoid using obvious correction methods to hide your typing errors. If you are mailing your price to the customer, you must make your presentation neat, well organized, attractive, professional, and convincing. With your computer, creating a neat presentation is easy—you can develop your own customized letterhead, add color if you have a color printer, and slip it into a nice binder that you can buy from your local office-supply company and you've got a first-rate proposal.

Phone Facts

A formal proposal by phone is not the way to present your bid. When you call in your price, people can't visualize what they are getting or the person giving them this information. They can't linger over the estimate and evaluate it. Chances are good that the customer may write down your price and then lose the piece of paper it was written on. Phone estimates also tend to make contractors look indifferent, since they don't even take the time to present a professional, written proposal.

Dress for Success

The dress code for contractors covers a wide range, depending on what you are selling and whom you are selling to. Whatever you wear, wear it well. Be neat and clean. Dress in a manner that you can be comfortable with. If you are uncomfortable in a three-piece suit, you will not project well to the customer. If you usually wear a uniform, wear it when presenting your proposal, but make sure it is clean and pressed. Jeans are acceptable and so are boots, but, again, both must be clean and neat. Avoid wearing tattered and stained clothing. It not only reflects on you but it also reflects on the product you wish to sell. You don't want the customer to be afraid to ask you sit on the furniture.

When you are deciding what to wear, consider the type of customer you will be meeting. If the customer is likely to be dressed casually, then you should dress casually. If you suspect a suit will fit in with your customer's attire, consider wearing a suit. Choose a wardrobe that will blend in with the customer. By overdressing, you might make the customer uneasy. And avoid flashy jewelry—that is usually a turn-off.

Your Vehicle

What you drive says a lot about you. If you drive an expensive car or truck, you are sending signals to the customer. When customers look out the window and see the contractor getting out of a fancy car, they think your prices will also be fancy. If you drive an old, dilapidated truck, they may assume that you are not very successful.

PRO POINTER

You will gain your own confidence through experience, but you must learn how to build confidence with your customers.

Choosing the best vehicle for your sales calls is a lot like choosing the proper clothing. You want a vehicle that will make the right statement. A clean van or pickup is fine for most any occasions. Cars are okay for sales calls, but avoid pulling up to your customer's house in an expensive luxury model unless you are catering to high-end clients.

Confidence, the Key to Success

Confidence is the key to success. You must be confident in yourself, and you must create confidence in the mind of the consumer. If you can get the customer's confidence, you can almost always get the sale.

Gaining the confidence of your customers can often be done by the way in which you talk to them. If you are able to sit down with customers and talk for an hour, the chances of getting the job increase greatly. Showing customers examples of your work can build confidence. Letters of reference from past customers will help in establishing trust. If you have the right personality and sales skills, you can create confidence by simply talking.

As a business owner you are also a sales professional, or at least you need to learn how to become one. Unless you hire outside sales staff, you are the person customers will be dealing with. If you learn basic sales skills, you will become a successful contractor even if your prices are higher than your competitors.

C H A P T E R F I F T E E N

Subcontractors, Suppliers, and Code Officials

Subcontractors, suppliers, and building officials are all players in the contracting businesses. When contractors know how to work with and control these participants in the business, their businesses prosper.

Hiring subcontractors who perform shoddy work will reflect directly on how the public views your company-as one that does not care about the quality of its product. When suppliers repeatedly fail to delivery quality products on time, your crews can come to an abrupt halt. And when you are ignorant of existing building codes, local officials will point out numerous violations, some of which can shut your job down. Code officers can also make your company look bad. When the work your company produces is cited for violations, your business loses credibility. In this chapter we will deal with each of these aspects of your business; let's start with subcontractors, who will share a large role in your day-to-day business activities.

PRO POINTER

If you want your business to be successful, you must develop strong relationships and learn how to work with subcontractors, suppliers, and building officials.

Subcontractors

As a contractor, it is common practice to hire subcontractors to perform various forms of work, typically site work, underground utilities, plumbing, HVAC, electrical, and often roofers and flooring contractors. The quality of these subcontractors is very important, since how they perform their work will have a direct reflection on your company. The average homeowner does not know who installed their plumbing work or who tiled the bathroom. They only know that it looks good, a reflection on you, or it looks substandard, also a reflection on you, the remodeler.

Subcontractors frequently have contact with your customers, and you should expect them to maintain the reputation you've worked hard to earn. However, without care in selecting a subcontractor and checking his work from time to time, you can expect some problems down the road. When you have a core group of dependable subs, you can respond to work quickly and efficiently. Customers love to get fast service, and subcontractors can give you this desirable dimension.

Subcontractors with good work habits and people skills will build your business. A lot of subs don't want all the responsibilities that go along with being the general contractor. If you select good, reputable subcontractors and treat them as you would your own employees, they will help you prosper as they prosper.

Suppliers and Vendors

You might not think that suppliers (or vendors, as they are often called) can have a major impact on your reputation, but they surely can. As a contractor, you are responsible for everything that happens on the job. If your supplier's delivery truck damages the customer's lawn, you're going to hear about it quickly, and the customer will look to you to repair the damage, even though you will expect the supplier to reimburse you for these expenses. When materials are not delivered on time, customers are not going to call the suppliers to complain; they are going to call you. As the general contractor, you take all the responsibilities and all the complaints.

PRO POINTER

If you want your customers to remain happy—and what contractor doesn't?—you must be in control of all aspects of the job, from getting the permit to doing the punch-out work and everything in between.

When you have reliable suppliers, they can improve your customer relations. When delivery drivers are courteous and professional, customers will appreciate it. When deliveries are made on time, customers are satisfied. Seeing to it that suppliers make and maintain good customer relations is up to you. You have to establish the rules for your suppliers to follow. If the suppliers are unwilling to abide by your rules,

SUBCONTRACTOR LIST

Service	Vendor	Phone	Date

FIGURE 15.1 Example of a list for subcontractors.

SUBCONTRACTOR SCHEDULE

Type of Service	Vendor Name	Phone Number	Date Scheduled

Notes/Changes:

FIGURE 15.2 Example of a subcontractor schedule form.

SUBCONTRACTOR SELECTION FORM

TYPE OF SERVICE	VENDOR NAME	PHONE NUMBER	DATE SCHEDULED
Site Work	N/A		
Footings	N/A		
Concrete	N/A		
Foundation	N/A		
Waterproofing	N/A		
Masonry	N/A		
Framing	J. P. Buildal	231-8294	7/3/04
Roofing	N/A		
Siding	N/A		
Exterior Trim	N/A		
Gutters	N/A		
Pest Control	N/A		
Plumbing/R-I	TMG Plumbing, Inc.	242-1987	7/9/04
HVAC/R-I	Warming's HVAC	379-9071	7/15/04
Electrical/R-I	Bright Electric	257-2225	7/18/04
Central Vacuum	N/A		
Insulation	Allstar Insulators	242-4792	7/24/04
Drywall	Hank's Drywall	379-6638	7/29/04
Painter	J. C. Brush	247-8931	8/15/04
Wallpaper	N/A		
Tile	N/A		
Cabinets	N/A		
Countertops	N/A		
Interior Trim	The Final Touch Co.	365-1962	8/8/04
Floor Covering	Carpet Magicians	483-8724	8/19/04
Plumbing/Final	Same	Same	8/21/04
HVAC/Final	Same	Same	8/22/04
Electrical/Final	Same	Same	8/23/04
Cleaning	N/A		
Paving	N/A		
Landscaping	N/A		

NOTES/CHANGES ______________________________

____

FIGURE 15.3 Example of a subcontractor selection form.

find new suppliers. Remember, in this case, you are the customer!

Materials

Delivery of the correct materials, in good shape and on time, can have a positive impact on your customer's peace of mind. If framing lumber, for example, shows up bowed with lots of splits or knotholes, your customer isn't going to be pleased. If the wrong materials are delivered and your crews are unable to continue working, you will lose time, and your customer will lose patience. Don't overlook the important role that materials play in the way customers view your business.

Choosing Your Product Lines

Choosing your product lines carefully is important to the success of your company. If you pick the wrong products, your business will fizzle. When you choose the proper products, ones that are familiar to your customers, they will sell themselves. As a business owner, you can use all the help you can get, so carry products that the public wants.

You can select products by doing some homework. Read magazines that appeal to the type of people you want as customers. For example, if you want to become known for the outstanding kitchen and bathroom designs used in your homes, read magazines that focus on kitchens and baths. Observe advertisements in the magazines.

PRO POINTER

By paying attention to ads, you will get a good idea of what your customers will be interested in.

Walk through the local stores that carry products you will be using or competing with. Take notes as you walk the aisles. Pay attention to what is on display and its cost. Take note of products that you see on television or advertised in the paper or in magazines. Your potential customers probably watch those same programs, read the same local paper, or subscribe to some of those magazines. This type of research will help you to target your product lines.

Visit some of your competitor's projects, particularly the successful ones. Go into their sample homes and check out the appliances that are installed, the type of flooring, the paint colors, the windows and doors. Look at homes under construction. See what your competition is doing. By simply riding past a construction site you can probably tell what types of doors, windows, siding, shingles, and similar items are being used.

MATERIAL ORDER LOG

SUPPLIER: ______________________________

____

____

DATE ORDER WAS PLACED: ______________________________

TIME ORDER WAS PLACED: ______________________________

NAME OF PERSON TAKING ORDER: ______________________________

PROMISED DELIVERY DATE: ______________________________

ORDER NUMBER: ______________________________

QUOTED PRICE: ______________________________

____

____

DATE OF FOLLOW-UP CALL: ______________________________

MANAGER'S NAME: ______________________________

TIME OF CALL TO MANAGER: ______________________________

MANAGER CONFIRMED DELIVERY DATE: ______________________________

MANAGER CONFIRMED PRICE: ______________________________

NOTES AND COMMENTS

____

____

____

____

____

____

____

____

____

____

____

FIGURE 15.4 Example of a material order log.

Your Company Name
Your Company Address
Your Company Phone and Fax Numbers

REJECTION OF GOODS

To: ____________________ Date: __________
____
____

We received goods from you under our order or contract dated ______________, 20 _____. However, we reject said goods for the reason(s) checked below:

___ Goods failed to be delivered within the required contract time.
___ Goods were defective or damaged as described on attached sheet.
___ Goods did not conform to sample or specifications as described on attached sheet.
___ Confirmation accepting our order, as required, has not been received, and we therefore ordered the goods from another supplier.
___ Prices for said goods do not conform to quote, catalog, or purchase order price.
___ Partial shipment only received; we do not accept partial shipments.
___ Other (please see attached sheet).

Please provide instructions for return of said goods at your expense. Rejection of said goods shall not waive any other claim we may have.

Sincerely,

____

Title: ____________________________

FIGURE 15.5 Example of a rejection of goods form.

This on-site investigating can put you in touch with what the public wants.

There are several ways in which you can conduct informal surveys. You can go door to door and do a cold-call canvassing of a neighborhood. You will experience a lot of rejection, but you will also get some answers. If you don't like knocking on doors, you can use a telephone. You can even have a computer make the phone calls and ask the questions for you. . Be sure to consult the national "Do Not Call" list.

PRO POINTER

When you talk to your customers, ask them about products that they'd like to see in their home. By doing so you not only get some valuable information but you give your customers the feeling that you really care about being up to date on current market trends.

You can also use direct mail. Direct mail is easy to target, and it's fast and effective. While mailing costs can be expensive, the results are often worth the expense. You could design a questionnaire to mail to potential customers. Done properly, your mailing will look like you have a sincere interest in what individuals want. It will appear this way because you will have a sincere interest.

Answers to your questionnaire will tell you what products to carry. You can improve the odds of having the pieces returned by self-addressing the response card. You can absorb the cost of the return postage by purchasing a permit from the local post office and having it printed on your cards. You can affix postage stamps to the cards, but this will cost more. With the permit from the post office, you pay only for the cards that are returned, not counting the permit fee. If you use postage stamps, you will paying for postage that may never be used.

To convince people to fill out your questionnaire, you need an incentive such as offering a discount from your normal fees. A better idea might be to make the questionnaire look more like a research project, which in fact it really is. If you design the piece to look like a respectable research effort, more people will respond to your questions and feel that you are a progressive businessperson interesting in learning what the public really wants.

Avoiding Delays in Material Deliveries

Avoiding delays in material deliveries is crucial to the success of your business. If your materials are delayed, your jobs will be delayed. If your jobs are delayed, your payments will be delayed. If your payments are delayed, your cash flow and credit history can falter, and if your cash flow slows down or disappears, your business is headed for trouble. How to avoid late material shipments? You can maintain acceptable delivery schedules by being involved personally.

Your job of maintaining the delivery schedule begins when you place the order. Some basic principles apply to keep your deliveries on schedule. First of all, place your orders far in advance of when you need them, thereby giving your supplier sufficient time to work it into the delivery schedule. Get the name of the person taking your order. Use a phone log to document all your calls. Ask the order taker to give written documentation for your delivery date. While you are at it, get the name of the store manager; you may need it if you have problems with the delivery. When you have the intended delivery date, stay on top of the delivery. If you have placed the order several days in advance, make follow-up phone calls to check the status of your material. Always get the names of the people you are talking to. You never know when you will have to lodge a complaint. Keep clear records of your dealings in an order log. By maintaining a telephone or personal presence you will establish a relationship with that supplier and the employees handling your deliveries. They will assume that if you are this attentive now, you will be tough to deal with if the order gets screwed up.

With a lot of effort and a little luck, your deliveries will be made on schedule. If the shipment does go astray, contact the store's manager. Advise the manager of the problem and the ripple effect it is creating for your business. Produce your documentation on the order. By showing the manager your written delivery date, employee names, and supporting documentation, such as material specifications, you will make a strong impression. This tactic will set you apart from the customers who complain but can't back up their complaint with facts. You will come across as a serious professional. If you don't get satisfactory results, move up the ladder to higher management or consider finding another source and let your current supplier know that you are looking around.

Choosing Subcontractors

Choosing subcontractors requires lots of time and effort. This is a key element in your business and requires a lot of consideration. Your business reputation will rise or fall based upon the quality of your subcontractors.

PHONE LOG

Date	Company Name	Contact Person	Remarks

FIGURE 15.6 Example of a phone log.

Your Company Name
Your Company Address
Your Company Phone and Fax Numbers

CANCEL ORDER

Date: ________________________

To: ________________________
____
____

I refer to our purchase order or contract dated ________, 20 ____, as attached.

Under said order, the goods were to be shipped by _____, 20 ___.

Because you failed to ship the goods within the required time, we cancel the order and reserve such further rights and remedies as we may have, including damage claims under the Uniform Commercial Code.

If said goods are in transit, they shall be refused and returned at your expense under your shipping instructions.

Sincerely,

____

Title: _______________________________________

FIGURE 15.7 Example of a cancel order form.

Your Company Name
Your Company Address
Your Company Phone and Fax Numbers

CANCELLATION OF BACKORDERED GOODS

Date: ______________________

To: ______________________
____
____

BE IT KNOWN, that pursuant to our purchase order dated ____________, 20 _____, as attached, we have received only a partial shipment. As noted by you on the packing invoice, some goods are out of stock or backordered.

Please be advised that we are canceling the backordered goods. Invoice us only for the goods received. If the backordered goods are in transit, please advise us at once and we shall give you further instructions.

Sincerely,

____

Title: ____________________________

FIGURE 15.8 Example of a cancellation of backordered goods form.

The Initial Contact

You will learn a lot about your subcontractors from the initial contact. Did they show up on time and have a neat physical appearance? If so, they're off to a good start. Your first impression may not be accurate, but you are sure to formulate one.

Not only will you form a first impression of the subcontractor, but they will also form an opinion of you, so you should plan to be free when they arrive at the appointed time. Make sure your office is somewhat neat and has an organized appearance. The potential importance of this meeting suggests that you should take control of the proceedings.

As a contractor, you will come to rely on subcontractors and you need to make a good impression. Have the topics you wish to discuss at hand and discuss them in a professional manner. Just as subcontractors may alienate you, you may alienate them. This, obviously, is not what you want to do. You want your first contact with subcontractors to be productive so you need to choreograph it carefully. Wasted time is potentially wasted money.

Just as you want to engage professional subcontractors, subcontractors want to work with successful contractors. One of subcontractors' greatest concerns is whether or not they will be paid promptly or even if they will be paid in full. If, as a general contractor, you come across as an unorganized, financially shaky business, subs will not be thrilled at the possibility of working with you.

Subcontractors and general contractors go together like peanut butter and jelly. If there is not a comfort level between the two parties, the working relationship will not succeed. Plain talk and honesty are the best traits to exhibit when talking to a subcontractor, and you will expect the same from them.

PRO POINTER

What can you do to attract quality subcontractors? If you project a professional image, one with a sound business philosophy, subcontractors will seek you out. Whether you are talking on the phone or in person, send the right messages. Let subcontractors know you are a professional and will accept nothing less from them.

Application Forms

Application forms can come in handy when interviewing subcontractors. While subs are not going to be traditional employees, it is not unreasonable to ask them to complete an employment application. The applications used may not resemble those used for employees, but you want to know as much about them as possible.

The application should contain questions pertaining to the type of work the subcontractor is equipped to perform. Asking for credit and work references is a reasonable request. Asking for their insurance coverage is crucial. If the subcontractor is not properly insured and one of the employees is injured on your job, you may be liable for those injuries. You can customize your applications to suit your needs. It may be wise to discuss the form and content of your subcontractor applications with an attorney.

Basic Interviews

There are many questions you will want to ask in the initial interview with a subcontractor. When you conduct your interviews, you want to derive as much insight into the qualities of the subcontractors as possible. These interviews will be the basis for your decision to hire or not to hire subcontractors.

If you have a professional office, that is the best place to meet subcontractors. If your office conditions don't reflect the image you want to give, meet the subcontractors on neutral ground. You could meet them in a coffee shop, restaurant, or other place. Pick a meeting place that will allow you to project your best image. One of the best places to meet your subcontractors is at their place of business. First of all, you'll impress them with the fact that you took the time to come to their office. You can ask for a tour of their shop to look at the way in which they store materials and equipment, possibly see one of their trucks being loaded to go on a job. In their office, you can talk to their bookkeeper, possibly their estimator, if they have one, and get a general idea of their office and field operations. During the interview, set the pace and go through your list of questions. Let the subcontractor talk, but you should set the pace for the interview.

Checking References

Checking references should be standard procedure when selecting subcontractors. If a subcontractor has been in business for any length of time, he or she should have references. Ask for these references, and follow up on them. But remember that the subcontractor is likely to give you only good references. As you talk to each reference, ask for other companies that the subcontractor has had a business relationship with and then contact them.

Your Company Name
Your Company Address
Your Company Phone and Fax Numbers

SUBCONTRACTOR QUESTIONNAIRE

Company name: ______________________________
Physical company address: ______________________________
Company mailing address: ______________________________
Company phone number: ______________________________
After-hours phone number: ______________________________
Company president/owner: ______________________________
President/owner address: ______________________________
President/owner phone number: ______________________________
How long has company been in business? ______________________________
Name of insurance company: ______________________________
Insurance company phone number: ______________________________
Does company have liability insurance? ______________________________
Amount of liability insurance coverage: ______________________________
Does company have worker's comp. insurance? ______________________________
Type of work company is licensed to do: ______________________________
List business or other license numbers: ______________________________
Where are licenses held? ______________________________
If applicable, are all workers licensed? ______________________________
Are there any lawsuits pending against the company? ______________________________
Has the company ever been sued? ______________________________
Does the company use subcontractors? ______________________________
Is the company bonded? ______________________________
With whom is the company bonded? ______________________________
Has the company had complaints filed against it? ______________________________
Are there any judgments against the company? ______________________________

FIGURE 15.9 Example of a subcontractor questionnaire.

Checking Credit

Another part of screening subcontractors is checking credit. By checking the credit ratings of subcontractors, you can determine a good deal about the individuals who own the company and some financial history of their business.

If subcontractors have had bad credit, it may not mean they should be rejected as a potential hire, since there may be extenuating circumstances. I've known a few subcontractors who were not paid by general contractors who skipped town, owing lots of people. As a result these two subs had difficulty paying their bills on time for quite a while. They finally absorbed these big losses and got back on track, but someone just looking at the bad credit report of a year or two ago without knowing all the facts would form the wrong opinion of these two reliable companies.

Credit reports can tell you a lot about the people you will be doing business with.

Read Between the Lines

In all your business endeavors you must learn to read between the lines. Credit reports are a good example of where the facts may not tell the whole story. Let's say you are reviewing a credit report and see that a subcontractor has filed for bankruptcy; would you subcontract work to this individual? Look further–you may be missing out on a good subcontractor. The fact that someone has filed for bankruptcy is not enough to rule out doing business with the individual. Individuals can get into financial trouble through no fault of their own, as the above example shows. You must be willing and able to decipher what you are seeing. When you learn to read between the lines, you will be a more effective businessperson.

Set the Guidelines

If you plan to utilize the services of the subcontractors you are interviewing, set the guidelines for doing business with your firm. If you require all your subcontractors to carry pagers, tell them so. If you require subcontractors to return your phone calls within an hour, make the point clearly. Remember that you are in control, but you can't expect people to read your mind. You have to let your desires be known.

Coming to Terms

Coming to terms is a key issue in selecting subcontractors. What you want and what the subcontractors want may not be the same. If you are going to do business with subcontractors, you should work out the terms of your working arrangements in advance.

Discussing contracts is a vital topic in meeting with subcontractors. You should go over your subcontract agreements with the subcontractors. If there are any questions or hesitations, resolve them in the meeting. You don't want to get into the middle of a job and find out that your subs will not play by the rules.

The more detail you go into in the early stages of your relationships, the more likely you are to develop good working conditions with your subcontractors. Just like your contracts with homeowners, you want your subcontractor agreements to be free of confusion and misinterpretation. Insert whatever clauses are appropriate to make sure subs understand what you expect. Take as much time as necessary, but remove any doubts from your contracts.

The contract should include the time and date when you expect the sub to be on the job. Include the time and date when their work is to be completed and accepted as satisfactory. If the sub fails to meet these dates, what leverage do you have? As far as starting date, you don't have much, but if he delays too much, you'll need to send notification that if he fails to start on such-and-such date, you will cancel the contract. This is the reason for having multiple subs in each trade ready to do business with you.

As far as completion is concerned, you have a little more leverage. A standard contract provision, known as a "performance clause," deals with this issue. A typical performance clause is similar to the following:

> The Subcontractor agrees to commence and complete the Subcontractor's work by ________________ and to perform the work at lesser or greater speeds, and in such quantities as , in the Contractor's judgment is required for the best progress of the job or as specifically requested by the Contractor. Should the subcontractor fail to prosecute the work ort any part thereof with promptness and diligence or fail to supply a sufficiency of skilled workers or materials of proper quality, the Contractor shall be liberty after seventy-two hours written notice to the Subcontractor to provide such labor and materials as may be necessary to complete the work and to deduct the cost and expense thereof from any money due or thereafter to become due to the Subcontractor under this agreement.

Another subject for the contract is job cleaning. Subcontractors are generally required to clean up their trash and debris, but many don't do it promptly, and many don't do it unless there is a clause in the contract such as:

> The Subcontractor shall at all times keep the job site clean and orderly and free from dirt arising out of the Subcontractor's work. At any time, upon the Contractor's request, Subcontractor shall immediately clean up and remove from the job site anything it is obligated to remove hereunder or Contractor may, at its discretion and without notice perform or cause to be performed such clean up and removal at the Subcontractor's expense.

Payment issues are always important. In the industry, there is widespread use of a "pay when paid" clause in many subcontract agreements. Some states have deemed these clauses invalid and you may not wish to use them. If you do, the clause typically states that the subcontractor will be paid within "X" days (generally 15 to 30 days) after the contractor has received payment for that work from the Owner. In effect, this pegs your payment to the subcontractor to payment from the owner. If you are paid promptly, your subcontractor will be as well. However, if the owner delays payment to you, you can delay that corresponding payment to your subcontractor.

Delays frequently occur in this business, and they often affect some of your subcontractors who share no responsibility in those delays. However a sub may look to the general contractor for reimbursement of costs they incurred because of these delays. These costs are referred to as "consequential damages." A clause in the subcontract agreement stipulating that "Contractor shall not be liable to Subcontractor for any damages or extra compensation that may occur from delays in performing work or furnishing materials or other causes attributable to other subcontractors, the owner, or any other persons" will do the trick.

Maintaining the Relationships

Once you hire new subcontractors, you must concentrate on maintaining the relationships. When you find good subs, you have probably invested a significant amount of time in their selection, and to avoid wasting this time you must work on building these relationships. This doesn't mean that you have to become buddies with your subs, but you do have to fulfill your commitments and expect them to fulfill theirs.

Your Company Name
Your Company Address
Your Company Phone and Fax Numbers

SUBCONTRACTOR AGREEMENT

This agreement, made this __________ day of __________, 20__, shall set forth the whole agreement, in its entirety, between Contractor and Subcontractor.

Contractor: ____, referred to herein as Contractor.

Job location: ____

Subcontractor: ____, referred to herein as Subcontractor.

The Contractor and Subcontractor agree to the following.

SCOPE OF WORK

Subcontractor shall perform all work as described below and provide all material to complete the work described below.

Subcontractor shall supply all labor and material to complete the work according to the attached plans and specifications. These attached plans and specifications have been initialed and signed by all parties. The work shall include, but is not limited to, the following: ____________________

____

____

____

____

COMMENCEMENT AND COMPLETION SCHEDULE

The work described above shall be started within _______ (___) days of verbal notice from Contractor, the projected start date is ___________. The Subcontractor shall complete the above work in a professional and expedient manner by no later than _______ (___) days from the start date. Time is of the essence in this contract. No extension of time will be valid without the Contractor's written consent. If Subcontractor does not complete the work in the time allowed, and if the lack of completion is not caused by the Contractor, the Subcontractor will be charged _______ ($______) dollars per day, for every day work extends beyond the completion date. This charge will be deducted from any payments due to the Subcontractor for work performed.

(Page 1 of 3. Please initial ________.)

FIGURE 15.10 Example of a subcontractor agreement. *(continued on next page)*

SUBCONTRACTOR AGREEMENT (continued)

CONTRACT SUM

The Contractor shall pay the Subcontractor for the performance of completed work subject to additions and deductions as authorized by this agreement or attached addendum. The contract sum is ____($__________).

PROGRESS PAYMENTS

The Contractor shall pay the Subcontractor installments as detailed below, once an acceptable insurance certificate has been filed by the Subcontractor with the Contractor. Contractor shall pay the Subcontractor as described:

____

All payments are subject to a site inspection and approval of work by the Contractor. Before final payment, the Subcontractor shall submit satisfactory evidence to the Contractor that no lien risk exists on the subject property.

WORKING CONDITIONS

Working hours will be _______ a.m. through _______ p.m., Monday through Friday. Subcontractor is required to clean work debris from the job site on a daily basis and leave the site in a clean and neat condition. Subcontractor shall be responsible for removal and disposal of all debris related to the job description.

CONTRACT ASSIGNMENT

Subcontractor shall not assign this contract or further subcontract the whole of this subcontract, without the written consent of the Contractor.

LAWS, PERMITS, FEES, AND NOTICES

Subcontractor shall be responsible for all required laws, permits, fees, or notices, required to perform the work stated herein.

WORK OF OTHERS

Subcontractor shall be responsible for any damage caused to existing conditions or other contractor's work. This damage will be repaired, and the Subcontractor charged for the expense and supervision of this work. The Subcontractor shall have the opportunity to quote a price for said repairs, but the Contractor is under no obligation to engage the Subcontractor to make said repairs. If a different subcontractor repairs the damage, the Subcontractor may be back charged for the cost of the repairs. Any repair costs will be deducted from any payments due to the Subcontractor. If no payments are due the Subcontractor, the Subcontractor shall pay the invoiced amount within _______ (_____) days.

(Page 2 of 3. Please initial ________.)

FIGURE 15.10 *(continued)* Example of a subcontractor agreement.

(continued on next page)

SUBCONTRACTOR AGREEMENT (continued)

WARRANTY

Subcontractor warrants to the Contractor, all work and materials for _________ from the final day of work performed.

INDEMNIFICATION

To the fullest extent allowed by law, the Subcontractor shall indemnify and hold harmless the Owner, the Contractor, and all of their agents and employees from and against all claims, damages, losses, and expenses.

This agreement, entered into on ___________________, 20____, shall constitute the whole agreement between Contractor and Subcontractor.

______________________________ ______________________________

Contractor Date Subcontractor Date

(Page 3 of 3)

Your Company Name
Your Company Address
Your Company Phone and Fax Numbers

SUBCONTRACTOR CONTRACT ADDENDUM

This addendum is an integral part of the contract dated _______________, between the Contractor, ______________________, and the Customer(s), ____, for the work being done on real estate commonly known as ____________.

The undersigned parties hereby agree to the following:

____

____

____

____

____

The above constitutes the only additions to the above-mentioned contract. No verbal agreements or other changes shall be valid unless made in writing and signed by all parties.

______________________________ ______________________________

Contractor Date Customer Date

____

Customer Date

FIGURE 15.10 *(continued)* Example of a subcontractor agreement.

If you tell a subcontractor that you will pay bills within five days of receiving them, you had better be prepared to pay the bills. When you agree to terms in your subcontractor agreements, stick to them. If you breach your agreements with subcontractors, they will, at some point decide to go elsewhere with their business.

Rating Subcontractors

Rating subcontractors will take some time and effort, but it will be worth it. You should rate subs in order to develop the best team possible. This rating procedure starts with the interview but doesn't stop there. You should look at many aspects of the subcontractor's business and history.

One of the first qualities you should evaluate is the work history of the subcontractor. Experience counts, not only in terms of the company but in the people you'll be working with. For example, a person with fifteen years of experience who has just gone into business may be a better choice than the one who has been in business for two years but only has five years of experience. There are advantages to choosing a subcontractor who has an established business.

If the subcontractor has been in business for awhile, there is a better chance that the business will last. New businesses often fail within the first two years, but businesses that have been around for between three to five years have a better chance of survival.

PRO POINTER

Business owners who survive these early years have business experience and dedication. These are the traits you should look for in a subcontractor.

The business practices employed by subcontractors can affect their desirability. This aspect may be difficult to assess in a single meeting, but by doing some research you can learn much about how the sub conducts business. You need a subcontractor who is responsive to your phone calls. With beepers and cell phones, quick communication is available to nearly everyone. Most subs will assure you that they are attentive to answering calls, but you need to verify their claims. A quick test can be conducted shortly after the subcontractor leaves your office. Call him. You know that he has just left the meeting and is not in his office, and you will find out how his phone is answered and how quickly he will return your call.

Before you commit to using a subcontractor, conduct a test. Call three contractors and schedule a bid meeting. Be prepared to discuss a pending job. If you don't

Your Company Name
Your Company Address
Your Company Phone and Fax Numbers

LETTER OF ENGAGEMENT

Client: ____
Street: ____
City/State/Zip: ____
Work phone: ______________ Home phone: ______________
Services requested: ____________________________________
____
____

Fee for services described above: $ _______________________
Payment to be made as follows: ___________________________
____

By signing this letter of engagement, you indicate your understanding that this engagement letter constitutes a contractual agreement between us for the services set forth. This engagement does not include any services not specifically stated in this letter. Additional services, which you may request, will be subject to separate arrangements, to be set forth in writing.

A representative of _______________________ has advised us that we should seek legal counsel prior to using information or materials received from ____
____

We the undersigned hereby release _________________, its employees, officers, shareholders, and representatives from any liability. We understand that we shall have no rights, claims, or recourse and waive any claims or rights we may have against __________________, its employees, officers, shareholders, and representatives. We further understand that we will pay all costs of collection of any amount due hereunder including reasonable attorney's fees.

_________________________ _________________________
Client Date Client Date

____
Company Representative Date

FIGURE 15.11 Example of a letter of engagement.

have any, create a simple project that you are going to ask them to bid. You will be looking for certain key responses. Will they be punctual in responding? How long will it take to get the quotes you want? Will their response be professional, clear in the scope of work included or excluded? Will they offer any suggestions to improve quality or price? If you created a "project," after you have received all their responses and the test is over, you can merely tell them that the job didn't go ahead, but you thank them for responding. But you have now found out how each subcontractor would have responded to a real situation.

Tools and equipment are another consideration when judging subcontractors. If your subcontractors don't have the necessary tools and equipment, they will not be able to give you the service you desire. Don't hesitate to inquire about the tools and equipment owned by the subcontractor.

Insurance coverage is a very important subject. You cannot afford to use subcontractors who are not properly insured. It is easy to lose your business in a lawsuit concerning personal or property damage. For your own protection, you must make sure your subcontractors carry the proper amounts of personal liability and property insurance. If an employee of the subcontractor is injured on your job and the sub doesn't have insurance, you could be liable for any claims that that employee may file. So for this and other reasons you need to have your subcontractors submit certificates of insurance to you *before* starting work on your project.

If the sub has employees other than close family members, workman's compensation insurance will be needed. Even if the subcontractor is not required to carry worker's comp, you should have a waiver signed by the business owner. The waiver, sometimes referred to as a "hold harmless" clause, should be prepared by your lawyer, so you can be protected from any potential claim.

PRO POINTER

If you use the services of subcontractors who are not properly insured, you may have to pay up at the end of the year. When your insurance company audits you, as they usually do, you will be responsible for paying penalties if you used improperly insured contractors. These penalties can amount to a substantial sum of money. To avoid losing money, make sure your subcontractors are currently insured for all necessary purposes.

Subcontractor Specialization

Many subcontractors specialize in various areas of work. Some HVAC contractors only fabricate ductwork. Others install

CONTRACTOR RATING SHEET

Job name: ______________ Date: ______________________

Category	Contractor 1	Contractor 2	Contractor 3
Contractor name			
Returns calls			
Licensed			
Insured			
Bonded			
References			
Price			
Experience			
Years in business			
Work quality			
Availability			
Deposit required			
Detailed quote			
Personality			
Punctual			
Gut reaction			

Notes: ____

FIGURE 15.12 Example of a contractor rating form.

CONTRACTOR COMPARISON SHEET

Category	Contractor 1	Contractor 2	Contractor 3
CONTRACTOR NAME			
RETURNS CALLS			
LICENSED			
INSURED			
BONDED			
REFERENCES			
PRICE			
EXPERIENCE			
YEARS IN BUSINESS			
WORK QUALITY			
AVAILABILITY			
DEPOSIT REQUIRED			
DETAILED QUOTE			
PERSONALITY			
PUNCTUAL			
GUT REACTION			

Notes: ______________________________

FIGURE 15.13 Example of a contractor comparison form.

equipment and others may only install low-voltage control wiring. Some electrical contractors specialize in low-voltage data and voice communication wiring and installations. When you are dealing with specialists, you may pay extra, but the end result may be a bargain. Remember that time is money, and when you save time you have a chance to make more money. With this in mind, ask potential subcontractors what they specialize in. You may find it cost-effective to use different subs for different jobs.

PRO POINTER

While specialists may charge higher fees, they are often worth the extra cost because they can complete their work more quickly, allowing you to bring in the next trade sooner.

Licenses

Licenses are another issue you should investigate when rating subcontractors. If subcontractors are not licensed as required by the local or state authorities, you can get into deep trouble. It is important that you only hire subcontractors that meet and have obtained standard licensing requirements. If you use unlicensed subcontractors, you are flirting with disaster.

Work Force

A subcontractor's work force is another consideration. How much work can the contractor handle? You don't want to hire a sub that cannot handle your workload. For this reason, you must know what the capabilities of the subcontractors are.

I've rarely heard a subcontractor refuse new work, even when he already has a full workload. Subcontractors who take on more work than they can handle will surely disappoint some builder—and it could be you. They will promise to be on the site, but, because of another contractor's demands that day, they'll go to the

PRO POINTER

If you give a small contractor too much work, you will find yourself in a bind. The small contractor may hire either temporary employees or inexperienced or substandard workers to keep your business.

other site. So you need to investigate how much work the subcontractor has and whether or not he can honestly service your job when needed.

While it is more convenient to work with a single subcontractor, it may not be practical due to his current workload, so you need to develop a core of subs in the same trade . As a safety precaution, you should have at least three subcontractors in each trade. This depth of subcontractors will give you more control.

It is also good to introduce competition among subcontractors. By using the same subcontractor over and over, other subs may be reluctant to bid on future work, assuming you always favor the one you have continued to hire, so you may not get truly competitive pricing.

Managing Subcontractors

Managing subcontractors will be much easier when you follow some simple rules. The most important rule is to document your dealings in writing. Other rules include:

- Create and use a subcontractor policy.
- Be professional and expect professionalism from the subs.
- Use written contracts with all of your subcontractors.
- Use change orders for all deviations in your agreement.
- Dictate start and finish dates in your agreement.
- Back charge subcontractors for costs due to poor performance or poor quality.
- Always have subs sign lien waivers when they are paid.
- Keep certificates of insurance on file for each sub.
- Don't allow extras unless they are agreed to in writing.
- Don't give advance contract deposits.@tx:
- Don't pay for work that hasn't been inspected.
- Use written instruments for all your business dealings.

Subcontractors can take advantage of you if you let them. However, if you establish and implement a strong subcontractor policy, you should be able to handle your subs. It is imperative that you remain in control. If subcontractors have the lead role, your company will be run by the subs not by you.

Dealing with Suppliers

Dealing with suppliers is not as simple as placing an order and waiting. Your business depends on the performance of suppliers, and it is up to you to set the pace for all your business dealings. Establish fair rules and make sure they follow them. Be firm but fair.

Establish a routine with your suppliers. If you are going to use purchase orders, use them with every order. When you want the job name and address written on your receipts, insist that they are always included. Are you going to allow employees to make purchases on your credit account? If so, set limits on how much can be purchased, and make sure everyone at the supply house knows which employees are authorized to charge on your account.

When you use a new supplier, make sure you understand the house rules. What is the return policy? Will you be charged a restocking fee? Will you get a discount if you pay your bill early? What is your discount percentage? Will the discount remain the same regardless of the volume you purchase? These are just some of the questions you should get answers to. If all goes well, you will be doing a lot of business with your suppliers. Since each of you depends on the other for making money, you should develop the best relationship possible.

PRO POINTER

Get to know the manager of the supply house. Without a doubt, there will come a time when you and the manager will have a problem to solve. At these times it helps to know each other.

Making Your Best Deal

How will you know when you are making your best deal? Is price the only consideration in the purchase of materials or the selection of subcontractors? No, service and quality are two other important factors in that equation. Getting the lowest price doesn't always mean you are getting the best deal. If you don't get quality and service to go along with a fair price, you are probably asking for trouble. Let me give you a few examples.

Assume that you have requested bids from five painters. You accept the lowest bid based on price alone. When the painters are scheduled to start the job, they don't show up. After calling and insisting that they be on the job by the next morning, you

Your Company Name
Your Company Address
Your Company Phone and Fax Numbers

PAYMENT ON SPECIFIC ACCOUNTS

Date: ________

To: ____________________
____
____

Our enclosed check number __________ for $____________
should be credited to the following charges or invoices only:

Invoice/Debt	Amount
	$

Payments herein shall be applied only to those specified items listed and shall not be applied, in whole or in part, to other obligations.

Sincerely,

____

Title: ______________

FIGURE 15.14 Example of a payment on specific accounts form.

get some satisfaction. The painters show up and start to work. You go back to your office, and at noon the homeowner calls, wanting to know where the painters are. You find out the painters left for a morning break and never came back.

This does not make your company look good in the eyes of your customer. What does the slowdown do to your cash flow? It crimps it, of course. You got the cheapest painter you could find, and your great deal doesn't look so good now. This type of problem is common, and you will have to do a better job of finding suitable subcontractors in the future.

For the next example, assume that you have ordered roof trusses from your supplier. After shopping prices, you decided to go with the lowest price, even though you had never dealt with the supplier before. The trusses are ordered and you are given a delivery date. All your work is scheduled around the delivery of the trusses.

If the trusses are to be used to replace a rotted roof structure, you can't tear off the old roof until you know the trusses are available. On the day of delivery, you call the supplier and inquire about the status of the trusses. You're told the trusses are on a delivery truck and will be on your job by mid-morning.

Your crew finishes framing work and is waiting to set trusses. It's nearly noon, and the trusses have not yet been delivered. Your crew is at a standstill. A phone call to the supplier reveals that the delivery truck broke down on the way to the job. (The broken truck ploy is an old one. Perhaps the truck did break down, but more likely the supplier messed up your delivery and is looking for an excuse.) You're told the trusses won't arrive until the next morning, but now what are you going to do? You've already lost money while your crew was idle waiting for the trusses

When you ask the supplier to transfer the trusses to a different delivery truck so you can get them immediately, you're told that the supplier doesn't have another truck capable of transporting the trusses. As it turns out, you have to leave the job until the trusses arrive, losing both time and money. And you have really upset your customer.

Would this have happened if you had used your regular supplier? Probably not, because that supplier has enough trucks to make a switch if necessary. Your great deal on inexpensive trusses has turned into a disaster. So, you see, price isn't everything.

Expediting Materials

Work cannot get done unless there are materials to work with. If there is no work there is no money coming in but lots of money going out. Since business owners are in business to make money, they need to keep materials available and flowing.

Placing an order and periodic follow-up calls takes a fair amount of time and effort, but if a worker has to leave the job to go pick up materials at a supply house, time and money are lost. Inaccurate take-offs and inattention to deliveries can cost contractors thousands of dollars. Is there anything that can be done to reduce these losses? Yes, by expediting materials, more time is saved and more money is made.

PRO POINTER

Learning how to expedite materials will keep your business running on the fast track.

All too many contractors call in a material order and forget about it. They don't make follow-up calls to check the status of the material. It is not until the material doesn't show up that these contractors take action. By then, time and money are being lost.

Many contractors never inventory materials when they are delivered. If 100 sheets of plywood were ordered, they assume that they received 100 sheets of plywood. Unfortunately, mistakes are frequently made with material deliveries. Quantities are not what they are supposed to be. Errors are made in the types of materials shipped. All these problems add up to more lost time and money.

When you place an order, have the order taker read the order back to you. Listen closely for mistakes. Call in advance to confirm delivery dates. If a supplier has forgotten to put you on the schedule, your phone call will correct the error before it becomes a problem.

When materials arrive, check the delivery for accuracy. Ideally, this should be done while the delivery driver is present. If you discover a fault with your order, call the supplier immediately. By catching blunders early, you can reduce your losses. And don't forget to note any deficiencies or discrepancies on the delivery ticket so there is a written record of the problem.

Keeping a log of material orders and delivery dates is one way of staying on top of your materials. One glance at the log will let you know the status of your orders. When you talk to various salespeople, record their names in your log. If there is a problem, it always helps to know whom you talked to last. Get a handle on your materials, and you will enjoy a more prosperous business.

Avoiding Common Problems

By avoiding common supplier and subcontractor problems, you can spend more time making money. Money is usually the major cause of disputes, and communication breakdowns cause the most confusion. If you can conquer these barriers, your business will be more enjoyable and more profitable.

PRO POINTER

The two biggest reasons for problems between contractors and suppliers or subcontractors are poor communication and money.

There are few excuses for problems in communication if you always use written agreements. When you give a subcontractor a spec sheet that calls for a specific make, model, and color, you eliminate confusion. If the subcontractor doesn't follow the written guidelines, an argument may ensue, but you will be the victor.

As for money, written documents can solve most of the problems caused by cash. When you have a written agreement that details a payment schedule, there is little for anyone to get upset with. By using written agreements, you can eliminate most of the causes for disagreements and arguments. It's a good idea to create a boilerplate bid form to use in conjunction with plans and specifications. When these forms are designed for specific trades, you can eliminate confusion and mistakes during the bidding process.

Building Good Relations with Building-Code Officials

Developing good relations with building-code officials is an important part of most contracting businesses. If your business depends upon inspections and approval by local or state building-code-enforcement officials, you will do well to get to know the inspectors.

Depending upon the type of construction work you are doing and the area in which you are working, rural community, suburb, or urban area, the number and type of building-inspection officials may vary. In small towns, one official may inspect many trades—framing, plumbing, electrical, HVAC—whereas in some large cities there could be an inspector for each of those trades.

These code officials are there to ensure the public that the applicable building codes have been met. Some builders may have a problem with these inspectors because they are not familiar with the codes and have subcontractors who don't really know or don't follow their code requirements. A builder who abides by the appropriate code will have little or no problem with the inspecting officials and in fact can learn a lot from them. Building credibility with these building-code officials is the key to getting along with them.

Your Company Name
Your Company Address
Your Company Phone and Fax Numbers

BID REQUEST

Contractor's name: ______________________________
Contractor's address: ______________________________
Contractor's city/state/zip: ______________________________
Contractor's phone number: ______________________________
Job location: ______________________________
Plans and specifications dated: ______________________________
Bid requested from: ______________________________
Type of work: ______________________________
Description of material to be quoted: ______________________________
____
____
____

All quotes to be based on attached plans and specifications. No substitutions allowed without written consent of customer.

Please provide quoted prices for the following: ______________________________
____
____
____
____
____
____

All bids must be submitted by: ______________________________

FIGURE 15.15 Example of a bid request form.

Your Company Name
Your Company Address
Your Company Phone and Fax Numbers

CODE VIOLATION NOTIFICATION

Contractor: ____

Contractor's address: ______________________________________

Contractor's city/state/zip: _________________________________

Contractor's phone number: _________________________________

Job location: ____

Date: ____

Type of work: ____

Subcontractor: ____

Address: ____

OFFICIAL NOTIFICATION OF CODE VIOLATIONS

On ________________, 20 _____ , I was notified by the local code enforcement officer of code violations in the work performed by your company. The violations must be corrected within _____ (___) business days, as per our contract dated ________________, 20 _____. Please contact the codes officer for a detailed explanation of the violations and required corrections. If the violations are not corrected within the allotted time, you may be penalized, as per our contract, for your actions in delaying the completion of this project. Thank you for your prompt attention to this matter.

______________________________ ______________________________

General Contractor Date

FIGURE 15.16 Example of a code violation notification.

CODE VIOLATION NOTIFICATION

CUSTOMER NAME: Mr. & Mrs. J. P. Homeowner
CUSTOMER ADDRESS: 192 Hometown Street
CUSTOMER CITY/STATE/ZIP: Ourtown, MO 00580
CUSTOMER PHONE NUMBER: (000) 555-1212
JOB LOCATION: Same
DATE: July 25, 2004
TYPE OF WORK: Electrical
CONTRACTOR: Flashy Electrical Service
ADDRESS: 689 Walnut Ridge, Boltz, MO 00580

OFFICIAL NOTIFICATION OF CODE VIOLATIONS

On July 24, 2004, I was notified by the local electrical code enforcement officer of code violations in the work performed by your company. The violations must be corrected within two business days, as per our contract dated July 1, 2004. Please contact the codes officer for a detailed explanation of the violations and required corrections. If the violations are not corrected within the allotted time, you may be penalized, as per our contract, for your actions, delaying the completion of this project. Thank you for your prompt attention to this matter.

______________________________ ____________________

J. P. Homeowner Date

FIGURE 15.17 Example of a code violation notification.

One of your goals must be avoiding rejected code-enforcement inspections. This goal is not difficult to achieve. Work gets rejected because it is not in compliance with the local code requirements. If you know and understand the code requirements, you shouldn't get many rejection slips. If you don't understand a portion of the code, consult with a code officer. It is part of an inspector's job to explain the code to you. When the work of your subcontractors is rejected, notify them, in writing, of their code violation.

PRO POINTER

Calling for inspections before they are ready is a sure way of aggravating a code official. Checking the job to make sure that everything looks O.K. and you are ready for your inspections is a sure way of building a relationship. Don't be afraid to smile and talk with your inspectors. By getting to know each other problems will be easier to resolve.

Again, attitude can have a bearing on the number of rejections you get. If you walk around with a chip on your shoulder, inspectors may look a little more closely for minute code infractions. If you play by the rules, you won't have much trouble with the officials. But don't ever try to put one over on a code official. If you get caught, your life on the job will be miserable for a long time to come. Inspectors can be a close-knit group. When you deceive one, others will get the word, and your work will be put under a microscope.

Learning to work well with subcontractors, suppliers, and building-code officials is essential to the success of a building business. Work hard to develop and keep good relationships. You depend on a team of people when you remodel homes for a living, so you need to concentrate on and develop team-building qualities.

CHAPTER SIXTEEN

Time Management Translates Into Stronger Earnings

Are you good at managing your money? Do you think you will be good at making money in your business? How about your time? Do you possess good time-management skills? Do you understand the principles behind time management? Well, if you are weak in either of these areas, you are going to have some trouble in your business. This chapter will help you strengthen your skills in both time and money management. Knowing how to make the most of your time will smooth the path to making more money.

Time is Money

It is a cliché, but time is money. Whether you bill your time on an hourly or a contract basis, lost time translates into lost money. To get the most out of your business, you have to get the most out of your time and money. How many times have you said you don't have time for this or that? Do you find yourself rushing around to get everything done, only to be frustrated that you didn't accomplish your goals? Poor time management is usually a factor in these circumstances.

PRO POINTER

There is a difference between not having time to accomplish a task and not taking the time to complete the job.

Let's look at a quick example. Consider yourself a busy business owner. Your morning starts early, and you work until the phone stops ringing at night. You know your day can include 12 to 14 hours of work. Obviously, time is a coveted commodity. There are only so many hours in the day. Even with all this work, getting out of the field and into the office has caused you to go a little soft. You want to work out, but when could you possibly manage to find the time?

This scenario is not unusual. Business owners are often obsessed with the advancement of their business. This obsession can ruin marriages, damage relationships with children, and drive the business owner to extreme behavioral swings. I always seem to be in high gear, but it is true that you need to allow some time for yourself.

PRO POINTER

If you are going to survive in the fast lane, you have to learn how to make pit stops. Otherwise, you are going to burn out.

Over the last 12 years I have changed my work habits considerably. I no longer go from 6:00 a.m. to 11:00 p.m. I spend time with my wife and children. I pursue some hobbies, and I'm going to start working out really soon, really I am. Okay, so some nights I write until 3:00 a.m., but I'm not missing time with my family. They are asleep. I wouldn't consider myself a workaholic, but I am very aggressive. In fact, if you want to win the business battle, you have to be aggressive. If you move too slowly in the rat race, the faster rats will run over you.

When you have something you want to do or should do, you can probably make time to do it. In the case of making time to work out, you can work out before work, at lunch, or after work. When you consider that most authorities say you only need twenty minutes of exercise every other day, that's not much time. You probably spend that much time reading the paper, drinking coffee, or thinking about what you are going to do on your day off if you ever take one. Time management is an individual act. Every individual will adjust to changes in his or her schedule differently. Let's see how you can budget your time more effectively.

Budget Your Time

Do you budget your time? Most people don't; they react rather than act. This is a major mistake for a business owner. If you compare it to a football game or boxing, it is easy to see. The team that scores first or the man who throws the first punch has the advantage. The other side has to react to this aggressive action. The opponent of the aggressive team or fighter doesn't have the luxury of planning the attack. He or she is too busy trying to thwart the offensive action of the first one out of the starting block. The same is true in business. Let me show you what I mean.

Consider this example. You are a remodeling contractor and you perform much of your own work. Your strongest competitor, another remodeler, has an outside sales staff and hires subcontractors, leaving more time to spend on business projections and evaluations. Both of you make about the same amount of money.

You try to be on the job by 7:30 a.m. and you rarely leave before 4:00 p.m. When you leave work after putting in a long day working with your tools, you then face the task of dealing with some estimates and other paperwork. By 6:30 p.m., you're home and having supper with your family. From 8:00 p.m. to 10:00 p.m. you are returning phone calls, setting up subcontractors, and paying bills, among other things. By midnight, you're in bed. This is a tough schedule, but not uncommon for someone with a small contracting business.

While you are spending long hours dealing with all field, office, and sales problems, your competitor seems to be gliding through life, in the office at 9:00 a.m. while you have been on the job since 6:30 a.m. Your competitor handles office work during the day while you burn the midnight candle.

PRO POINTER

Even if you are happy with 18-hour days, unless you learn how to budget your time and free up some moments when you can step back to take a look at how things are going and how some things might be done better, you will find it difficult to grow your business.

Doing business your way, you probably have fewer on-the-job problems because you are on the job. You are not paying out money to commissioned salespeople, and you are in total control of your business, as much as anyone ever is. You are basically married to your business, and while you have chosen to create a working life, your competitor has created a business. Which would you prefer to have? Your competitor has his business running smoothly because he has good management and has made good management decisions. Part of this management is time management.

I started out by having a working life. Now I have a business. This is not to say that I don't work. I probably work more hours than most business owners, but I enjoy most of what I do. For me, work is not drudgery. I use subcontractors, sell my own work, do some of my own fieldwork, and do my own photography, writing, and consulting. I have diversified and have variety in my work life, and this variety actually helps to make me more productive and keeps me on an even keel. However, to accomplish this goal, I've had to perfect my time-management skills. If you want to shape your business, you will also need to learn how to use time management.

Know When You Are Wasting Your Time

It is easy to get caught up in the heat of the battle and lose your objectivity. If this happens, you will not realize that you are working harder and losing ground. To run a successful business, you must be able to step back and look at the business operation from an objective point of view.

Taking an unbiased look at your operational procedures can be difficult. Many of my clients hire me to troubleshoot their businesses for just this reason. They are too close to the situation to see their problems—too close to the forest to see the trees. However, you don't have to hire an outside consultant to evaluate your business. You can do it yourself.

PRO POINTER

For effective time management, you must know when you are wasting your time

To determine how much time you are wasting, you may have to take some time off from your day-to-day activities and try to determine where those hours go. One of the best ways to pinpoint the ways you spend your time during a workday is to prepare a time log. Keep a log of everything you do and the time you spend on each activity. This log will show you where you have been spending your time—where you have been spending it productively and where you have been wasting time.

Your log can be written or dictated into a tape recorder, whichever you are most comfortable with. As soon as you wake up, start your log. Keep track of everything you do from the time you wake up until you retire for the evening. This should include not only your business activity but also your personal functions.

I know it may be a burden at times to stop and record or write down what you are doing, but you must discipline yourself to make entries in your log with every activity you undertake. Whether it is brushing your teeth, walking the dog, going to the mailbox, or making business calls, enter your actions in the log. If you don't take the log seriously, this experiment will not work.

Keep your log for at least two weeks. At the end of that time, review the log page by page. Scrutinize your entries for possible wasted time. If you find you talk for more than five minutes during your business calls, take a close look at what you are talking about. Certainly, there are times when business calls deserve thirty minutes or more, but most calls can be accomplished in five minutes or less.

Look for little aspects of your daily life that could be changed. Do you sit at the breakfast table and read the paper? How long do you spend reading the paper? Does this reading help you in your business? If your reading doesn't pertain to your business, maybe you should consider curtailing the time you spend at the table. If you derive pleasure from reading the paper, it's not a bad thing to do. However, if you are caught up in a habit of reading the paper for half an hour and wouldn't feel deprived without this time, you have just found time for your workout.

PRO POINTER

You will probably find many red flags as you go down your time log. Almost everyone has little habits that rob time. Most of these routine activities bear little importance in the daily life of the people, but they persist because they are habits. Before you can change your bad habits, you must identify them. A time log will help you to expose your wasted time.

Controlling Long-Winded Gab Sessions

Employees are supposed to make money for you, but they can drain you of your profits. If you spend too much time talking with employees, you are losing money in two ways. You are not free to do your job, and your employees are talking to you; they're not working. Let me give you a case history that will drive this point home.

PRO POINTER

Part of your time-management routine should include controlling long-winded gab sessions with employees.

During one of my consulting assignments, I was working with a service company to see how it could improve its efficiency. The business owner was smart. He knew that if I was brought in and introduced as a consultant, the employees would not act as they normally did. For this reason, I was introduced as just another employee.

I acted like an employee, dressed like an employee, and got to know the employees. The first week I worked for the company, I found a major loss of income—the result of upper-management conversations with employees.

When the crews would come into the office for their daily assignments, it was common for them to hang around for half an hour talking to the management. The talk was not business-related. There were ten employees, an operations manager, and an office manager involved in the conversations. This company was charging $35 per hour for its labor rate. Based on billable time, the ten employees wasting thirty minutes a day were costing the business owner $175 a day. The office manager and the operations manager were making a combined income in excess of $45,000 a year. When you tally up the cost to the business owner for these morning talks, the annual lost income was in the neighborhood of $45,500. What seemed like a simple, friendly morning talk was actually a business-threatening cash loss.

As you continue in business, you will find that suppliers of materials and equipment will begin to call you on the phone or stop by the job site to talk about their product. They have all the time in the world to stop and talk you to about Monday Night Football or the American League Playoffs, but don't fall into the habit of spending a lot of time with these salespeople. Don't shoo them away, because they can bring you lots of information about new products, sale items, even how your competitors might be making out. But it is also easy to spend time discussing things that are not job-related, and you have to discipline yourself to cut the conversation short. The same would apply to phone calls from vendors at a time when you are busy. Politely tell them you are busy; you would like to talk to them (if in fact you do), but they should call back later—the next day, after 4:00, or whatever is a more convenient time.

PRO POINTER

If you set your appointments to achieve maximum efficiency, you will find additional free time in your day. You can convert this extra time into money. If you prefer, you can spend the time you save with your hobbies or spend the additional time with your friends or family. In any event, setting efficient appointments will give you more time to use for the purpose of your choice.

Set Your Appointments for Maximum Efficiency

How do you set efficient appointments? You can arrange them in a logical order. It is beneficial to schedule meetings in your office instead of a customer's home or office. By meeting in your office, you save travel time. However, if you are having a sales meeting, you might be better off to sacrifice some time and meet with the potential customer on his home turf.

People are more comfortable in their own home or office. When you are trying to sell a customer a remodeling job, you want the customer to be as comfortable as possible. But, for now, we are dealing with time management, so let's concentrate on why you want the meetings to convene in your office.

By setting appointments in your office, you save time and gain control. You save time, because you can work until the person you are meeting arrives. How many times have you gone on appointments only to have the other party be late or never show up? Have you ever thought of how much time these tardy or broken appointments cost you? Schedule appointments in your office whenever possible. If your client is late, you can continue to work. If the appointment is broken, you haven't lost any time from work.

Did you know that people will feel more intimidated when there is a desk between you and them? Are you aware of the signals you are sending when you put your hands behind your head and lean back in your chair? Body language says a lot, and if you investigate sales techniques, you will discover how your body language will influence your meetings. Keeping a desk between yourself and the person you are meeting is intimidating. In fact, you will often be in a more advantageous position to move out in front of your desk. People will be more receptive and less intimidated. Leaning back in your chair with your hands behind your head signals that you are in control of the conversation. Body language is a powerful sales tool, and good sales people know how to read the signals people are sending with their physical movements.

PRO POINTER

There is a side benefit to meeting people in your office. You will be more at ease, and the people you are meeting will feel at a disadvantage. While this can be detrimental in a sales meeting, it will work to your advantage when negotiating with suppliers and subcontractors. Under these circumstances, you have the control.

Reduce Lost Time in the Office

To maximize your time, you must reduce lost time in the office. It is common to think that if you are in the office you're working, but this theory doesn't always work. A lot of people sit in the office without accomplishing any productive work, and they lose time and money.

Where do you lose time in the office? Do you sit in the office for hours at a time, waiting for the phone to ring? If you do, get a cellular phone or an answering service, and get out there looking for business. If your phone isn't bringing you business, you must go out after it.

Can you type? Do you take an hour to type a single proposal? If you do, it might be worth your while to find a typist who is an independent contractor. Hire the typist to transcribe your voice tapes into neatly typed pages. This eliminates work you are not good at and allows you to do what you do best.

Assess your office skills and rate them. If filing is not your strong suit, find someone to do your filing for you. If you have an aversion to talking on the telephone, hire someone with an excellent phone presence to take your calls. You can use a checklist to rate your in-office performance and concentrate on what you need to work on.

The A List and the B List

Creating a daily "To Do" list for the next day is always helpful. It makes you think about what you have to do, which might involve making telephone calls the night before. So you accomplish two things: listing the important items you need to attend to tomorrow and listing phone calls you have to make tonight in order to make these things happen tomorrow.

There should be two lists: the A List consists of items that <u>must</u> be done and should therefore be relatively short; the B List includes items that, once the A list is complete, you will attack. This list can be a little longer because items not completed can be rolled over to the next day.

When you find that you can't complete all the A List items, you may need to rearrange your priorities. Did all these items absolutely, positively have to be completed today? You'll find the answer may be "Well, no." The A List should include only those items of work that you must complete. This approach will also help you establish your priorities and sort out the really important things from those that can wait a day or two.

Reduce Lost Time in the Field

If you reduce lost time in the field, you will make more money. Reducing lost time in the field is similar to reducing time in the office. You can use the same type of checklist to appraise your performance. However, the ways for improving the quality of the time you spend in the field will require some different tactics.

Whether you are doing your own work or supervising subcontractors, you can spend most of your time in the field. There is nothing wrong with being out of the office as long as you are being productive. But how productive are you?

There are many ways to reduce the time you lose in the field. A mobile phone can make a huge difference in your in-field production. Working with a tape recorder will give you some advantages. Setting up a filing system in your work vehicle will help you to stay organized and save time. Let's take a look at the wonders a tape recorder can work for you.

Using a Tape Recorder to Improve Efficiency

You may be amazed by how the use of a tape recorder can improve your efficiency. Micro cassette tape recorders can boost your productivity quite a bit. Many contractors spend an enormous amount of time driving. Tape recorders that can be purchased for as little as $25 can turn this previously wasted driving time into productive time. Letters can be dictated and notes and marketing ideas recorded. The device in effect becomes a portable notebook. Upon returning to your office or while at home, you can review the day's notes and jot down the important ones. You can remove the micro cassette, write the date, time, and subject matter on it, and store it away in case you need to refer to it in the future. Slip in a new cassette and you're ready for the next day. Tape recorders are definitely worth strong consideration.

Should You Have a Cell Phone?

Whether or not you should buy a cell phone depends on you and your business. Cellular phones can be a boom to your business. Almost any owner of a service business can benefit from a cell phone. Wireless phone service has become a very competitive business, and there are lots of deals being advertised for plans for a large number of airtime hours for one basic monthly rate or free cell phones in return for a one-year contract for phone service. But before you sign up, ask for the hidden charges—an "advertised" monthly rate of $29.95 can balloon to $37 or more once the other "fees," taxes, and other costs are added. And be sure that the cost per minute for calls in excess of the monthly allowance doesn't send your phone bill to astronomical heights. Read any cell-phone contract carefully to avoid a surprise after that first month's bill arrives.

The potential benefits of being able to communicate on the move are many. A general contractor can call from a remote job site and cancel a concrete delivery if a job has failed its footing inspection. If the delivery is not stopped in time, the contractor will have to pay for the concrete even though it's not used. And on the other hand if you need a load of concrete but you are far from a landline phone, your cell phone can get the material on site when you need it.

PRO POINTER

The cost of a cell phone can be a bargain. If a $5 call results in a $10,000 job, you've made a wise decision.

Other benefits can accrue when service calls are involved. A service technician can call ahead to make sure the next homeowner on the schedule is home and ready for service. If the homeowner has forgotten the appointment, the service technician can move up the schedule to the next service call. This can save at least an hour's labor charge. When you add up this type of savings over a year's time, the amount can be substantial.

If you have a cell phone and get stuck in traffic, you can call ahead and advise your party that you are on your way but are stuck in traffic and will be late–you'll call back as some as traffic starts moving again.. By keeping your customers and business associates aware of your schedule, you will have less broken appointments and disappointed clients.

If you have on-the-road communication, you can always be reached for emergency or highly sensitive issues. When customers know they can reach you at any time, they are more comfortable doing business with you.

Time truly is money. You might benefit from sitting in on one of the many seminars available on the subject of time management. Learning how to budget your time is just as important as knowing how to budget your money. After all, your time is what you're selling. By honing your time-management skills, you will be a better contractor and your odds of survival will be much higher.

PRO POINTER

With the progress in modern technology and the competitive nature of service businesses, I think cellular phones are really a necessity.

C H A P T E R S E V E N T E E N

Customer Satisfaction

Customers and public relations are two concerns that every business owner must consider. Without customers, a business is worthless. Without good public relations, customers will be lost. These two pieces of the business puzzle go together; you can't have one without the other.

Contractors certainly recognize the need for customers, but many underestimate the importance of public relations. By being unaware of the importance of public relations, many contractors will ultimately lose business and they may not even be aware of the reasons why they have, only that their business volume is down.

If you don't want to fall into that trap, you need to work on your public-relation skills. When you improve these skills, you will see an improvement in your business. The improvement may not be obvious, but if you look closely, you will see it.

Meeting Your Customers on Their Level

Dealing with customers on their level is important. You want those people to be as comfortable as possible. Sometimes just a change in clothes or a slightly different approach is all that's necessary. Being able to assume the personality of a customer makes it easier to close a deal. But above all, be yourself. People can see right through a phony.

We talked about clothes and cars before, but it might be worth repeating. When you are dealing with customers, attempt to fit into what the customers are comfortable with. This might mean wearing a suit in the morning and jeans in the afternoon. You might find it advantageous to switch from your family car to your pick-up truck depending upon that the situation calls for.

PRO POINTER

The most successful sales people are the ones who are can change their sales approach when dealing with different types of customers.

Without Customers, Your Business Isn't Worth Much

Customers are what give your business value. How you treat existing customers can influence the long-range success of your business. If you alienate customers, they won't give you return business or referrals. It is much less expensive to keep good customers than it is to find them. Once you have established a customer base, work hard to keep it.

Qualifying Your Customers

When qualifying your customers, don't be afraid to ask questions. Qualifying customers includes a number of issues. One of the first questions you may be concerned with is the customer's ability to pay for the work being considered. This may seem like a basic concern of any businessman, but quite often this subject is not discussed when talking to a prospective customer.

Some customers may not pay their bills on time or have been known not to pay because of real or imagined complaints they raise when bills are presented to them. The reasons for nonpayment are extensive, but the end result is not getting paid. No business owner can afford to work for free. Let's look at some of the reasons you may not get paid by your customers.

PRO POINTER

While it is true that the customer is hiring you, you have the right to know a little about who you are working for.

Loan Denial

Denial of a loan may cause a well-meaning customer to be unable to pay your fees. If you are doing a large remodeling job or room addition, big ticket items, your customer will usually be depending on borrowed funds and, if the loan isn't approved, there will probably be insufficient funds available to proceed with the project.

This may not seem worth considering like a high risk, but many contractors put themselves into positions to take a direct hit because they started work before their customer's loan was approved. Contractors, especially new contractors, are often anxious to work. When a contract is signed for a big job, they start the job immediately, before a loan has been arranged and approved. This is bad business. You may feel awkward requesting proof of available funds, but you will feel worse if you don't get paid.

PRO POINTER

Don't be reluctant to ask your customers to show evidence of the money required to do the job. This could amount to looking at a loan agreement or your customer's bank statement.

Deadbeats

There will always be deadbeats in all walks of life; professionals at beating honest people out of their hard-earned money. Avoiding deadbeats can be difficult because they are not always easy to identify. They come in all shapes and sizes. To protect yourself from this group of customers, get their permission and run a credit check on them. This is good business for all of your customers. You never know when the sweetest, most trusting person is going to turn out to be a bad debt.

Your Fault

Sometimes nonpayment will be your fault. If you have not made the customer happy, you could have trouble collecting your cash. When you qualify your customers, try to read them for trouble signs. If you feel friction in the meeting, perhaps you should pass on the job and look for another customer.

Some customers are unnecessarily picky and will delight in telling you how they badgered their washer-dryer salesman until their new machines were replaced or how they have been unhappy with the painter who painted their apartment. You will be the next one on their list and, if they bad mouth other salespeople or tradesmen, they'll probably do the same to you. So do you really want this job that could turn into a nightmare?

Death

Death is always a pretty good excuse for not paying bills. While you can't avoid a customer's demise, you can make arrangements in your contract to cover the death contingency. Ask your attorney to draft a clause that will hold the heirs and estate responsible for your fees, if the client passes away.

Bankruptcy

If a customer owes you money and files for bankruptcy protection, your chances of being paid are all but nonexistent. During the qualifying stage you can screen the customer's credit rating and financial strength. If the customer is financially healthy when you start the job, there is limited risk of losing your money in the bankruptcy courts. Look for credit card debt. It is one of the leading causes of personal bankruptcies because high interest rates cause an initial debt or debts to increase dramatically when unpaid balances pile up.

How to Satisfy Your Customers

Do you know how to satisfy your customers? The answer to this question is complex. Not all customers are alike, and what will satisfy one customer may infuriate another. You must look at each customer individually. There will, of course, be similarities between customers, but each person will have at least a slightly different opinion of what is required of you.

There are some basic principles to follow when working with customers. They are:

- Keep your promises.
- Return phone calls promptly.
- Maintain an open and honest relationship.
- Be punctual.
- Listen to the requests of your customers.
- Do quality work.

- Don't overcharge consumers.
- Stand behind your work.
- Be professional at all times.
- Don't take your customers for granted.
- Prioritize your warranty work.
- When feasible, give customers what they want.
- If possible, give customers more than they expect.

If you follow those basic rules, you should have a high ratio of happy customers. There will, of course, be some customers you can't satisfy. When encounter this type of client, grin and bear it.

If you deal with an unreasonable person, end the relationship as quickly and professionally as you can. Avoid name-calling and arguments. If the customer is dead wrong, defend your position calmly, but if the circumstances are questionable, cut your losses and get out of the game. Avoiding conflicts and striving for customer satisfaction will do you much more good than standing on a soapbox screaming to the world what a jerk someone is.

PRO POINTER

Even if customers are irrational at times, go to extremes to please them. One angry customer will spread more bad word-of-mouth advertising than ten satisfied customers, and you don't want this type of publicity.

Learn When to Give and When to Take

A business relationship is similar to a marriage. In any relationship, there is a certain amount of give and take required. If you are hard-headed and bullish, you may win the battle, only to lose the war. For success in business, you have to learn the fine art of flexibility and, at times, compromise.

If you are too considerate and giving, some of the people in this world will take advantage of you. If you are cold and take a strong stand without wavering or giving in, you will probably lose some business.

PRO POINTER

A good definition of compromise is when each party to a negotiation walks away feeling unhappy.

To have a harmonious business, you must learn to blend give and take into a masterful mix.

It's funny, in a way. People go into business to be their own boss, but as a business owner, you have more bosses than ever before. Every customer you serve is your boss. Being in business is not the easy life some people think it is. Owning a business is hard work, and sometimes you have to do things you don't want to.

Learning to compromise, especially on money matters, can be a sobering experience. But, if you are going to stay in business, you are going to have to make compromises. The key is learning the difference between compromise and giving in.

When with a customer, talking it out is a good course of action. Actually, at times, listening may be a better course of action. If you listen carefully to what your customer is complaining about, you can often find a simple solution to the problem.

PRO POINTER

Becoming a good listener is one of the best ways to keep your customers happy. Once you have heard what the customer has to say, propose a reasonable solution. You may not come to terms on the first attempt, but if you follow this procedure, you have a good chance of finding common ground.

Public Relation Skills—Essential in Service Businesses

Public relation skills are essential in service businesses. To make your business successful, you need the support of the public and you will win over the public by demonstrating strong public relation skills.

You can start learning about pubic relations by reading books or you can take college courses and attend seminars. If you don't want to take the time to go to classes or read books about public relations, you can buy cassette tapes.. How you learn the skills is not the issue, the fact that you learn the skills is what's important.

Public relations is a career field in itself. There are people that specialize in this kind of work. You probably will not be able to master these skills overnight, But that's okay, you don't have to learn the skills all at once. In fact, most people refine their skills over years of work experience.

Every time you deal with the public, you can build on your public relation skills. This earn-while-you-learn process may not be the fastest way to develop good people skills, but it is one of the best. Reading and going to classes will give you the basics, but you have to deal with people to put those basics into motion. Study the principles used in public relations and refine them as you go.

Establishing Clear Communication Channels with Customers

Establishing clear communication channels with customers is essential in making your business better. If you and your customers understand each other, you will have far fewer problems. Your first contact with customers will often come over the phone. In this first contact, you will not be able to read body language and facial expressions. What is said and the way it is said is what you will have to judge each other. For this reason, it is important to speak clearly and project a cheerful, upbeat attitude.

Once past the phone conversation, a face-to-face meeting will likely occur. During this meeting, pay attention to the words and actions of your prospective customers. Also, be selective in your conversation. Four-letter words may have their place on the job but definitely have no place in customer relations. If you smoke, refrain from doing so.

When you start to discuss estimates and contracts, keep your thoughts concise and prepare a written estimate and contract so that you can review them with the customer. If the customer wants to have the documents reviewed by an attorney, by all means, let them.

You want to establish a comfort level for the customer. By keeping your contracts clear and easy to read, you will keep your customers satisfied. Clear communication is essential to good business.

PRO POINTER

If you watch and listen to the people you are meeting with, you will learn how to handle the potential clients.

Who in Your Organization Should Deal with the Customers?

Are you the best person to meet with clients? Do you have the required skills to sell the job? Obviously, no one can be the best at every aspect of running a business. Smart business owners know their strengths and their limitations.

You have to evaluate your personal skills. Being a great mechanic does not make you great at clerical tasks. Just because you can hammer a nail or hang a door doesn't mean you can sit down at a computer and be an instant wizard on the keyboard. You are probably better at some tasks than others.

Once you decide where you could use some help, work towards getting that help. This doesn't necessarily mean you have to hire employees. You can farm much of your work out to independent contractors such as commissioned salespeople, independent bookkeepers, and other specialists. Using skilled independent professionals can make your business more efficient. While your employees or independent contractors are doing their jobs, you can devote more of your time to doing what you do best.

PRO POINTER

To maximize your business potential, you need to list and examine your strong points and weaknesses. And be honest about them.

How to Defuse Tense Situations

Learn how to defuse tense situations that arise during your contact with customers. It may take some time but will be well worth it. If you choose to turn away from difficult situations, you will have a tough time in business. Business deals have the potential to turn into tense situations and it doesn't take much to make a good deal turn sour. Many times jobs will run smoothly until the last few days of the project.

PRO POINTER

For some reason, the end of the job seems to be the most difficult part.

When you have a job nearly complete, you have spent a lot of time, money, and energy to keep the customer happy. If you lose control at the end of the job, all of your previous efforts will be wasted. Normally, talking through a bad time can solve the problem. Sometimes just being a good listener is enough to resolve disputes. Some people may yell and scream to get the anger out of their system and, if that happens, let them expend their anger before you respond.

Most on-the-job confrontations don't elevate to extremes. Generally, work-related disputes can be resolved with open communication and occasionally, compromise. Every angry customer will react a little differently. It may be necessary to step away from the argument and look at the real reasons why the blow-up has occurred and be honest with yourself. Do you share a little or a lot of the blame? The best you can do is remain calm and reasonable.

Calming a Disgruntled Customer

Calming a disgruntled customer is similar to defusing a tense situation, but it may not always be the same. A customer can be dissatisfied without showing any signs of dissatisfaction. In fact, the calm customer who is displeased can be more difficult to read and to work with than the customer who is shouting. When you have an unhappy customer on your hands, you have to find a remedy that will appease the client. If you are able to carry on a normal conversation with the consumer, resolving the problem shouldn't be too difficult.

Ask your customer why they are dissatisfied and before responding, consider the other person's position. Do they have a legitimate gripe? If they do, take action to rectify the situation. If you disagree with the customer's opinion, discuss the problem in more depth. First of all, put the customer at ease. Assure them that you are willing to be reasonable, but that you need more facts to understand their position fully. By starting this way, the customer should remain calm and business-like. On the other hand, if you open your defense aggressively, the situation may escalate to a tense and unpleasant shouting match.

PRO POINTER

It is usually best to attempt a settlement of disputes in a relaxed atmosphere. If your crews are banging hammers and buzzing saws, ask the customer if the two of you can find a more suitable place to discuss your differences. This accomplishes two goals you get the customer in a congenial setting and your workers will not witness the disagreement.

Ask the client to repeat their grievance. Pay close attention, and see if the story remains the same. If the customer has additional complaints or varies from the initial comments, you may have some trouble. This behavior would indicate a person that is hard to please. If the complaint is essentially the same as it was when you first heard it, you have a good chance of resolving the problem amicably.

Think before you speak. If you are good at thinking on your feet, you will do better than people who need time before arriving at a conclusion. Your customer is going to expect answers now, not next week. But if the issue is complex, tell them you'll need a little time to think about resolving the issue. Tell them you'll get back to them in a day or two and make sure you do. Once you know what you want to say, say it sincerely and with conviction. Let the customer know you believe strongly in your position, but that you are willing to compromise.

You may find that you and your customer will exchange several opinions and offers before reaching an amicable decision. When you both agree on a plan, put the plan in by writing a change order. You will seal the deal and reduce the risk of having to negotiate it further at a later date.

Building a Reference List from Existing Customers

As you know, building a reference list from existing customers is an excellent way to produce new business. By asking your customers to give a letter of reference or to complete a performance-rating card, you are accumulating a valuable stack of ready references.

Asking customers for the names and phone numbers of friends or relatives is another way to get new business. Ask your customers for the names and phone numbers of anyone who might be interested in your professional services and then call them introducing yourself as the contractor working on Mrs. Smith's house. By knowing the neighbor's name, you have a good chance of starting a conversation and keeping the conversation going. You may be surprised at how many of the people your present customer refers you to might be interested in your services.

On-the-Job Decisions Can Be Dangerous

Making on-the-job decisions can be dangerous if they are snap decisions. These types of decisions can often lead to mistakes that rum the gamut from upsetting the homeowner to causing extra work for yourself. When possible, give yourself some time to consider your decisions, before they are made. Of course, there will be times when you have to make quick decisions but try to avoid them when you can.

PRO POINTER

There are two types of contractor-related liens. A mechanic's lien is a lien that may be placed by people who have not been paid for labor provided on a job. A materialman's lien is a lien levied by suppliers that have not been paid for materials supplied to a job.

Lien Rights And Waivers

Lien rights and waivers can have a significant impact on your business. These factors can work for you or against you. A lien right is the right to place a lien against property you have supplied labor or material for and not been paid for. What is a lien waiver? A lien waiver is a legal document when signed, relinquishes your lien rights, but does not relinquish your right to sue.

Lien-right laws vary from jurisdiction to jurisdiction, but they exist to protect workers and suppliers. Lien rights and the format for lien waivers may vary from state to state, so to learn more about these topics, consult with your attorney.

As a contractor, you will probably be asked to sign lien waivers and you will most likely ask subcontractors to sign lien waivers. When a job is being financed, the lender will often require lien waivers to be signed for every cash disbursement. When the person being paid signs a lien waiver, that person gives up their right to lien the property, for either unpaid labor or material. This protects the property owner and the lender.

Some contractors are approved for short-form lien waivers and others must use long-form lien waivers. A short-form lien waiver is a form that only the general contractor signs. In signing a short-form waiver, the contractor is attesting to the fact that all subcontractors have been paid for work done to a certain point.

The contractors approved for short-form lien waivers have usually been in business for awhile and have strong company assets. Property owners and lenders are taking a bigger risk in allowing contractors to sign short-form waivers. If the general contractor signs the waiver, but has not paid the subcontractors or suppliers, the unpaid parties may still lien the property. The general contractor will be responsible for having the lien removed, but there is some risk that the general contractor will not have the funds to settle the issue and have the lien removed.

PRO POINTER

As a general contractor, it is good business to have subcontractors and suppliers sign lien waivers, even if a lender or property owner does not require one.

Long-form lien waivers are more time consuming for contractors, but the property owners and lenders are in a much safer position. With a long-form lien waiver, anyone providing labor or materials for a job must sign the lien waiver at the time of payment. This way, no one can say they weren't paid. Their signature will be on the lien waiver.

Having your vendors sign off on waivers when they are paid, insures that liens will not be placed against the properties you are working on. This is just one more paperwork step that can help you avoid conflicts and trouble.

Solidifying Agreed-Upon Plans and Specifications

Solidifying agreed-upon plans and specifications is a step that should be done before starting a job. Once you and your customer agree on a set of plans and specifications, you should memorialize the agreement.

When you have completed the final contract, go over each page with the owner and when you are presenting a final set of plans and specifications, do the same thing. Remember that most people can't read prints, so you'll have the added task of explaining them. Point out the bathroom and kitchen areas, where the cabinets, fixtures and appliances are located. Explain wall sections, door and window details and so forth. Not only will you eliminate some future misinterpretations but you will impress your customer with your knowledge of construction details. When this review is over have your customers sign each page. Date the documents and make notations that the documents are the final and working plans and specifications. Further note that no changes will be made to the documents unless all parties agree to the changes in a written change order. When your customers sign below these notes, you have a solid set of plans and specifications. This procedure will eliminate the possibility of the customers coming to you later and saying that the job is not in compliance with the plans and specs. If this were to happen, all you would have to do is produce the signed documents and show that your work is in accordance with the agreed-upon plans and specifications.

The value of signed agreements is very important. Clear contracts and supporting documents will prevent ambiguous arguments. Your signed documentation will prove who is in the right. Keeping customers happy is not too difficult if you have a good working system. Most contractors can enjoy friendly relationships with their customers throughout the course of long jobs. To succeed in keeping customers happy, you may have to go out of your way at times, but the end result is worthwhile.

C H A P T E R   E I G H T E E N

Growing Your Business From Year to Year

Planning for your future is one of the first things you should do as a new business owner. This job will entail thinking about insurance, benefits, and retirement plans, all of which can be very perplexing. These areas of your business are not simple, and the responsibilities for you as a business owner are much more imposing than they were as an employee.

As a business owner, when you think of insurance, you must consider all aspects of the issue. Take health insurance, for example, not only the cost but the coverage. When you were working for someone else, the cost for insurance was deducted from your weekly paycheck. Now that you are the owner of your company you have lots of insurance questions to ask and to answer. You must now establish your own insurance program, pay all the costs, consider tax consequences, and determine what impact employees will have on the program you choose.

PRO POINTER

If you have employees or plan to hire them, benefit packages are a serious consideration. If you want to attract and keep the best employees, it has become standard practice for employers to provide their workers with benefits.

The task of establishing and administering benefit packages can get complicated. There are many options available, each with its own advantages and disadvantages.

If you are not aware of the legal implications concerning employee benefit plans, you can into serious trouble. Whether you are looking at retirement plans for yourself or your employees, the possibilities can be mind-boggling. Setting up a plan for yourself is one thing, establishing programs for employees is another.

Many business owners are unaware of the options available for insurance, benefits, and retirement plans. Now that you are an employer not an employee, the responsibility is on your shoulders, and it can be quite a burden. If you are the only employee of your company, your choices are easier. However, if you hire others to work for your company, you may have some studying to do. This chapter is going to prepare you for the kinds of decisions you will have to make when planning for the future of your business and your own long-range requirements.

Company-Provided Insurance for Yourself

Putting an insurance program in place for yourself, when no other employees are involved, is not difficult. However, choosing the right plan will take some research. Will the cost of your insurance be a deductible expense? If your business is structured as a corporation, you should be able to get some tax relief from the insurance premiums the company pays. If your business is not a corporation, you may not be able to deduct all, or a portion, of your personal insurance premiums. With medical costs skyrocketing, health insurance is considered one of the most important forms of insurance today. Dental insurance is not as critical as health insurance, but it does provide some additional peace of mind. Disability insurance is often ignored, but it can become very important if you are injured or contract a severe illness. Life insurance may not be too important if you are single, but it takes on added importance when you have a family.

Key-person insurance may not be critical for a small business as long as there is enough life insurance in force. Let's take a closer look at each of these types of coverage and see how they fit into your business plans.

Health Insurance

Health insurance is expensive and the plans are complex. Deciding on the type of insurance to get will require research and thought. Let's look at some of the things to consider when you are researching health insurance. They are:

- Pre-existing conditions
- Deductibles

- Limits of coverage
- Waiting period before insurance coverage begins
- Co-pays
- Dependent coverage

Pre-existing conditions may present a problem when applying for health insurance. For example, if you have problems with your back when you apply for a new policy, the insurance company may refuse to cover medical expenses related to these back problems. If you have had a pregnancy that involved surgery or medical attention beyond the normal child-birth requirements, a reoccurrence of these circumstances may not be covered by your new policy. It is possible to obtain insurance coverage where pre-existing conditions are not eliminated from coverage but the cost may be prohibitive. Maybe the protection is worth the additional cost.

The deductible payments for insurance plans vary. Typically, the higher the deductible , the lower your monthly premiums.. A plan with a $200 deductible will cost more on a monthly basis than a plan with a $500 deductible. It is wise to choose a plan with a higher deductible and lower installment payments.

PRO POINTER

Most insurance companies will not cover expenses related to a pre-existing condition.

Limits of Standard Coverage

Before you buy any insurance plan, know what the limits of standard coverage are. Not all policies cover all possible circumstances. Read policies closely and ask questions. The insurance company may not have to disclose facts to you unless they are asked direct questions.

Waiting Period

It is possible that an insurance policy will require a waiting period. These waiting periods stipulate that a specific amount of time must pass before a procedure is covered. For example, most insurance would not cover the costs of a pregnancy until after a waiting period has passed. Since the insured may have been pregnant when the policy was taken out, the waiting period eliminates the risk to the insurance company. Determine if the policy you are considering has a waiting period and if so, what conditions apply.

Co-Payments

Most health plans call for the insured to make co-payments. This means that you will be responsible for paying a portion of your medical expenses. A common co-payment amount is twenty percent of the costs incurred. You pay twenty percent and the insurance company pays eighty percent. However, the split on how much each party pays can vary. You might find that you are responsible for thirty percent of the bills.

Some coverages are more generous and pay nearly the entire cost of your medical expenses. For example, you may only pay a few dollars for each office visit to your doctor. These pay-all policies cost more, but they provide excellent coverage and don't require any out-of-pocket cash.

Dependent Coverage

If you have dependents, you will be interested in dependent coverage. Will your dependents receive the same coverage as you? Will the premiums be at reduced rates for the additional coverage? Are there limits on dependent coverage? Is there an age limit on the coverage extended to your dependents? All of these are questions you should ask about dependent coverage.

Rate Increases

Rate increases are almost automatic on an annual basis and vary from insurer to insurer. Inquire about the insurance company's rate increase policy. Will you face increases quarterly, semi-annually, or annually? Inquire about caps on the amount of increase at any one period. For example, if your rates increase on an annual basis, how much is the maximum increase? With insurance, you can never ask too many questions.

Group Advantages

As a business owner, you may be eligible for a group advantage. Some insurance companies will take several small groups of customers and create one large group. This type of grouping is designed to offer coverage at lower rates. Normally, your company will need at least two employees for this type of coverage, but the savings may be worth putting your spouse on the payroll. Check with your insurance representative for the requirements of joining a group plan.

Dental Insurance

Dental insurance is valuable for people with bad teeth. If you have paid for crowns or root canals lately, you know they aren't cheap. Should you buy dental insurance?

The decision is yours, but dental insurance can be well worth its cost for the right people. When you shop for dental insurance, you can ask about the same questions you ask for health insurance. Like health insurance, dental insurance comes in may forms. Choosing the right policy will be a matter of personal needs.

If you decide to buy dental insurance, expect to go through a waiting period for major-expense coverage. While some policies include routine maintenance immediately, you may have to wait for those more expensive crowns and caps. The waiting period for major work is usually one year. Many dental plans will pay no more than one half of your major expenses. For example, if you are getting an $800 crown, your insurance might only pay $400.

Disability Insurance

Disability insurance provides protection against lost income due to injuries and sudden disabling illnesses. Short-term disability policies are designed to provide assistance for a short period of time. Long-term disability will continue to make payments for an extended period of time.

Disability policies provide you with a percentage of your normal income while you are unable to work. The percentage of your income paid will depend on your policy. These policies may also include a pre-existing condition waiver. Let me give you examples of how each type of disability plan works.

Short-Term Disability

Short-term disability polices will set a limit on the amount of time you can receive benefits. Six months is a common benchmark for the maximum period of time you may collect from a short-term policy.

There is usually a short waiting period before the disability income (DI) will becomes effective In most cases, you will have to be out of work for at least a week before you can collect on your DI. The amount you can collect will be a percentage of your normal income. A plan that pays up to fifty percent of your income is not unusual. However, there are generally limits on the maximum amount you can collect.

For example, your policy may pay fifty percent of your normal weekly pay, but it might stipulate that the maximum you can receive in any given week is $300. Obviously, if you make more than $600. a week, and most contractors do, you will not be getting half of your income in benefits.

Long-Term Disability

Long-term disability works on a similar principle as short-term disability. These plans may pay a higher percentage of your income than short-term DI. There will be limits on the minimum and the maximum monthly payments, but the length of time you can collect payments is frequently unlimited.

Life Insurance

Life insurance doesn't seem very important until you have dependents. As a single person there may be little need to worry about how people will get along without your income when you die. But marriage and a family will change all that. In case of death there will be bills to pay and people to provide for. When you have family, life insurance becomes important.

How much life insurance do you need? What type of life insurance will suit your needs best? These are the two common life insurance questions every person should ask themselves. Life insurance must be tailored to your personal requirements.

How Much Life Insurance Is Enough?

The amount of life insurance coverage you need will depend on several factors. The first factor is the number of your dependents. A person with only a spouse will need less insurance than a person with a spouse and two children. Another factor is your income.

PRO POINTER

If you are leaving behind a spouse and children, the spouse may not have the earning ability to support the remaining family members. So you should carry enough insurance to allow for investments and long-term support.

Many people suggest buying insurance coverage based on a multiple of your annual income while others say insurance benefits should equal your annual salary. More people are inclined to believe it is better to have coverage equal to three year's worth of income. Your spouse's employment conditions may influence this decision.

If your spouse isn't working and hasn't worked for some time, it may be difficult finding a job. If you have been the sole provider, it will take time for them to establish a new life. Can all of this take place in one year? It could, but it would be a strain. So, if your insurance proceeds equal one year's income your spouse is under extra pressure, and don't forget, there will be burial expenses and other related expenses to be paid out of the benefits you bequeath.

Is your spouse capable of being self-supporting? If you died today, how many personal and business debts would you leave your family? Consider this when making your decision about the amount of life insurance you need.

As you can see, there are numerous factors to consider in determining the face amount of your life insurance. Some people look at life insurance as a one-time payment to the bereaved family. They assume that leaving their spouse $100,000, in cash, is more than adequate. In this mindset, the spouse is expected to live off the $100,000 until a new life is built. This isn't a bad plan, but there is another aspect to consider.

Term Life Insurance

Term life insurance is one of the least expensive forms of life insurance you can buy. While it is the least expensive, it may not be the best value. There are many types of term policies. Some of the programs include premiums that increase annually and other policies have face amounts that are reduced each year. Some may do both.

Term insurance is fine as a supplemental life insurance, but it may not be the best choice as a primary policy in your prime earning years while building assets. Term policies can protect your family from incurring your debts. For example, if you are buying a house with a thirty-year mortgage and die while there is a substantial loan outstanding, what will your spouse do? If the spouse can't afford the house payments alone, the house will have to be sold. However, if you have term life insurance, the proceeds from the policy could be used to satisfy the mortgage on the house. As time passes, the amount you owe on the home is reduced, so the reducing term insurance is not such a bad deal. You are paying only for the insurance you need, while you need it.

PRO POINTER

If you depend on term life insurance as your only life insurance, you will discover as you grow older, the premiums will go up and the value will go down. If you live a normal life span, the policy may not be worth much at the time of your death.

Whole-Life Policies

Whole-life policies are more expensive than term insurance, but they are more attractive because the face amount of these policies doesn't decrease and the premiums don't go up.

There are other advantages to a whole-life policy. As you make your monthly payments, you are building a cash-value in the policy. In effect, you are creating a savings account, of a sort. Later in life, if you need some quick cash, you can borrow against your built-up cash-value. Interest rates on these loans are usually very low, and you can pay back the money at your discretion.

If you reach a point in life where you no longer want to maintain your life insurance, you can cash a whole-life policy in and receive the money from the cash-value. These policies are considered one of the best available for the long term.

Universal and Variable Policies

Universal and variable life policies are variations of whole-life policies. These policies feature investment opportunities for your premium dollars. As you pay your premiums, you are building cash-value and your account is earning interest. The interest you are earning is rolled-over and is not taxable, unless it is withdrawn. Many business owners choose these policies.

Key-Person Insurance

If you have been in business, you have probably heard of key-person insurance. This is a form of life insurance that protects a company against the death of a vital employee. The employee is insured by the employer and the employer pays the insurance premiums. If the employee dies, the proceeds of the insurance goes to the employing company. This allows the company to have a cash buffer until the key employee can be replaced. Unless you are in a partnership or a corporation with other stockholders, you shouldn't need key-man insurance. Regular life insurance can protect your family and cover your business debts.

PRO POINTER

If you have a partner that you depend upon heavily, you might want to set up a key-person plan.

Other Options for Life Insurance

There are many other options for life insurance and these options can be overwhelming. There are many riders that can be added to standard policies, and terms and conditions can be adjusted to meet every conceivable need. Due to the complexity of insurance programs, you should talk to several insurance professionals before making a buying decision.

Choosing an Insurance Company

Choosing an insurance company is no easy job, but it can be one of the most critical aspects of your insurance planning. No one wants to pay premiums on insurance for years only to have the insurance company go out of business. Not all insurance companies have the same financial strength. The investment abilities of some companies are much better than those of other companies.

Choose your insurance firm carefully. Research the company and attempt to establish its financial power and track record. By talking with your state agencies and going to major libraries, you should be able to find performance ratings on the various companies.

PRO POINTER

Dig deep into a company's background before you depend on them to protect you.

Employee Benefits

PRO POINTER

As an employer, you are responsible for the compliance with federal and state laws and regulations. If you fail to execute your duties in accordance with those laws and regulations, you can wind up in serious trouble.

Benefits for your employees can be even harder to decipher than your own. There are a number of federal and state laws and regulations governing benefits to employees and you should consult with an expert in that field before you finalize your plans. The benefits you offer to your employees may include any of the insurance coverages we have already discussed. However, when you are setting up plans for employees, you may have to follow some additional guidelines.

Most companies use an employment manual to explain company polices to their employees. These policy manuals tell the employees what benefits they may be eligible for and when their eligibility begins. It is important that you treat all of your employees equally. You should not provide benefits for your key employees and deny the same offering to your other employees. If you do this, you are asking for trouble. The policy manual makes it easy for you to set and maintain protocol.

As you are shopping for benefit plans to offer your employees, you will find a host of them. Every company offering benefit services will tell you their benefits are the best. Wading through the myriad possibilities will take some time.

Many employers choose an insurance company that offers multiple benefits in a single plan. The benefits can include coverage for medical, dental, life, disability, and accident insurance. This type of employee package can be costly, but it is an attractive feature when you are trying to hire and keep top-notch employees.

Some of these plans allow employees to make some of their own choices in the types of coverage they want. The employer gives each employee a set allowance to allocate to various types of coverage, then the employee is free to customize his or her individual plan. This type of employee package is often referred to as a flexible benefit package and is sometimes called a cafeteria plan, a plan that allows the employees the ability to choose from various benefits. Other benefits you might offer your employees include paid sick leave, paid personal days, paid vacation, retirement plans, and bonus programs. Retirement plans for you and your employees will be discussed in detail later in the chapter.

Before you make a decision about providing benefits to your employees, research the rules and regulations you must follow. Talk to your attorney, your insurance agent, and your state agencies and you should be able to obtain all the information needed to stay on the right track.

Making Plans for Your Later Years

You should start making plans for your later years now. There is no time like the present to prepare for the future. In planning your future, you must define the paths you want to take with your business. Will your business be handed down to your children? Will the business be run by an employee when you retire? Are you interested in selling your business at some point in the future? These are only some of the questions you should start asking yourself now.

PRO POINTER

If you have hopes of one day giving your business, or the management of it, to your children, discuss it with them as soon as possible. The sooner the kids become involved in the business, the better they will be prepared to handle the responsibilities when you step down.

Passing The Business Along to Your Children

Passing the business along to your children is a fine way to keep your company going when you are thinking about retirement.

However, some children will have no desire to own or operate the business you spent years building. They may just have their own dreams to fulfill and they don't include your business.

Don't count on your children being overly enthusiastic about taking the reins, and don't become angry with them when they want to pursue other goals. After all, you wanted to build your own business, maybe they want the same freedom. Taking over the family business can put a lot of strain on devoted children. If they do to well, you might be offended that they are more capable in business than you were. If they perform poorly, they will feel they have let you down. Respect the wishes of your children and maintain a unified family.

Allowing Employees to Manage Your Business

Allowing employees to manage your business can be a hard pill to swallow, even when it's only for a few days. Would you trust them to mind the store while you took a vacation? Does the thought of having someone else at the helm of your business send shivers down your back? Putting your business into the hands of employees may take some getting used to. When you're comfortable with their performance under your watchful eye, take a short vacation.

PRO POINTER

If your plans call for having employees manage your business, start testing the waters now. Delegate duties to your best people and see how they handle them.

You will never know how the managers will function under pressure until you let them take control. If you are standing behind them every step of the way, they may be nervous and not perform to their best abilities. By being too close at hand, the managers may rely on you to make the tough calls. Get away from the business and let them have a go at it. If something does go wrong, you can step back in and pick up the pieces quickly. This is the only way you are going to be able to assess fully the abilities of your chosen few.

If while you're gone the business runs smoothly, give the managers a little more authority. Keep testing the employees with additional responsibilities. If you have the right people, you will be able to enjoy life more and rest comfortably, knowing you have good people to back you up.

Grooming Your Business for Sale

Grooming your business for sale is an important step towards liquidation. If you wake up one morning and decide to sell your business immediately, you are going to make mistakes. When your long-range plans call for the sale of the business, begin your preparations early. When the time comes to put the business on the auction block, you will be ready to make your best deal.

Closing the Doors

Closing the doors to a business you have invested your life in can be traumatic. You will probably feel you are throwing away a part of yourself. If shutting down is the ultimate fate for your business, prepare yourself mentally for the final days.

Reducing Your Workload

As an alternative to closing the doors, you might consider reducing your workload. Going into semi-retirement might be the ideal answer to your problems. You can be selective in the work you do, and you can enjoy some additional income. This option is very appealing to a lot of contractors. Again, proper advance planning is the key to making your desires reality.

Liability Insurance

Liability insurance is one type of insurance coverage no business can afford to be without. The extent of coverage needed will vary, but all business ventures should be protected with liability insurance. Most business owners are aware of what liability coverage is and their need for it. However, some readers may not be familiar with this type of insurance. For that reason, please allow me to explain how liability insurance works.

General liability insurance protects its holder from claims arising from personal injury or property damage. When a company has a general liability policy, all representatives of the company are typically covered under the policy, while performing company business. The cost of liability insurance will be determined by the nature of your business. Rates will be lower for someone engaged in relatively safe endeavors compared to those assessed against businesses dealing in high-risk ventures. For example, if you own a blasting company and work with explosives, your premiums will be higher than those who install interior trim molding.

PRO POINTER

Without adequate coverage against liability claims, you could lose your business and all of your other assets. Contractors are in particular need of this type of insurance. With so many possibilities for accidents on the job site, you can't afford to do business without it.

Worker's Compensation Insurance

Worker's compensation insurance is insurance that is generally required by individual states for companies having employees that are not close family members. The cost of this insurance can be very high, but it is a necessity for most businesses with non-family employees. Worker's comp insurance benefits your employees. If employees are injured in the performance of their duties on your payroll, this insurance will help them financially. The employees may receive payment for their medical expenses that are related to the injury. If employees are disabled, they may receive partial disability income from the program. Other events, such as a fatal injury, could result in similar benefits to the heirs of the employees.

The cost of worker's compensation insurance are based on the company's total payroll expenses and the types of work performed by various employees. The rate for a secretary will be much lower than the rate for a roofer. Each employee is put into a job classification and rated for a degree of risk. Once the risk of injury and other factors are assessed, an estimated premium is established.

At the end of the year, the insurance carrier will conduct an audit of the insured company's payroll expenses. It will be determined how much was actually paid out in payroll and to what job classifications the wages were paid. At this time, the insurance company will render an accurate accounting of what is owed or due to the company. Since some preliminary annual estimates are high, it is possible a company will receive a refund. However, if the original estimate was low, the company must pay the additional premium requirements.

PRO POINTER

Inattention to safety on the job may cause you to lose the services of your most productive people for long periods of time, and excessive accidents will increase your workers comp insurance rates.

Worker's compensation insurance premiums are based upon the company's safety record: a good one and rates are low; a high one and rates are high. In most states it takes three years of good safety records to erase one bad year. So poor safety will saddle you with increased insurance costs for three years. That's another reason why a company safety plan is important.

Worker's Comp for Subcontractors

Worker's comp for subcontractors can give you a nasty surprise. When you engage a subcontractor to work for your company, you might be held responsible for the cost of workmen's compensation insurance on that sub. You can avoid this by requiring subcontractors to furnish you with their certificate of insurance before allowing them to do any work.

The certificate of insurance will come from the issuing insurance company. When you receive the certificate of insurance, check it for coverage and expiration. When you are satisfied that the sub has proper insurance, file the certificate for future proof of insurance.

When your insurance company audits you at the end of the year, you may need to produce certificates of insurance on all of your independent contractors. You cannot afford to let your guard down on this one. Paying premiums for insurance that subcontractors should be responsible for will cost you.

> **PRO POINTER**
>
> Don't accept a copy of an insurance certificate that a subcontractor hands you. The policy may not still be in force.

Some contractors deduct money from payments due subcontractors when the subs don't carry the necessary insurance. The money is used at the end of the year when the contractors must settle up with their insurance companies. While this has been done for years, I don't recommend it. It is best to require the subcontractors to carry and provide proof of their own insurance.

Retirement Plan Options

There are lots of different retirement plan options. Whether you're looking for a plan for yourself or a plan for your employees, you will have many to choose from. Let's look at some of the most common methods of building retirement capital.

Rental Properties

Rental properties can be an ideal source of retirement income for yourself. Real estate is one of the best ways to keep up with the rising rates of inflation. Inflation is one of your biggest enemies when planning for retirement. With some investments, the money earned from the investment will not be eaten up by inflation. Real estate has the edge in these circumstances, because of its typical pattern of appreciation.

Rental real estate can be advantageous to you now and later. When you first buy income properties, the net rental income may not turn a profit for you, but the tax advantages can be significant. Tax laws change frequently and you need to check with your accountant to determine the latest IRS rulings pertaining to investment properties.

To make the most of your tax advantages, you must maintain an active interest in the management of your rental properties. If you are merely a passive investor, you will miss out on the bulk of the tax savings. However, being a contractor, you should be well suited towards being a landlord. You have the ability or the contacts to keep maintenance costs at a minimum.

If you own rental property that breaks even, you're doing fine in your retirement plans. While you are not turning a profit, you are paying off the real estate. If you begin your real estate investing early, when you retire you will be able to count on a steady income. Income-producing real estate allows you to win three ways. The first way is in the form of routine cash-flow. The rents you collect will allow you to have some spending money.

Rental properties help you increase your net worth. As your buildings become paid for, you gain equity. This equity can be used as leverage to borrow money against other properties. When you have enough equity in rental real estate, you can literally live on borrowed money for the rest of your life.

The third option you have with real estate is selling it. As you have been paying off the mortgages over the years, your real estate should have increased in value. If you don't want to be an active landlord in your later years, you can sell the property for a profit and live off the proceeds. Since you will be selling the real estate at current prices, you will not be losing ground to the effects of inflation.

If you have the temperament and time to be a landlord, rental properties are one of the best retirement plans you can come up with. However, you will have some problems along the way. Running rental properties can be a thankless job. Tenants are not always the easiest people to get along with, and there are many laws and regulations you must comply with. Some people simply are not cut out to be property managers. If you don't think real estate is your way to retirement riches, let's examine some other options.

Keogh Plans

Keogh plans for self-employed people can get a little complicated. If you are self-employed, you can contribute up to twenty-five percent of your earnings to the fund. However, the maximum dollar contribution is capped at $30,000. This formula gets a little trickier.

The twenty-five percent you are allowed to contribute is not computed on your gross earnings alone. When you determine how much you are going to fund your Keogh with, you must subtract that amount from your earnings. Then, you may contribute up to twenty-five percent of what is left of your earnings. In effect, you can only fund twenty-percent of your total earnings. This is a little complicated. Let me give you an example.

Let's say you had a great year and earned $100,000. You want to contribute $20,000 to your retirement plan. After subtracting the $20,000 from your earnings, you are left with $80,000. Twenty-five percent of $80,000 is $20,000, the maximum you can invest.

For older people, setting up a defined-benefit plan can allow larger contributions. However, these plans are expensive to establish and maintain. If you have employees, these plans become even more confusing. As the employer, you not only must deduct the contribution to your personal plan before arriving at the earnings-figure used to factor your maximum contribution, you must also deduct the contributions you made as your part of the employees' contributions. There are other rules that apply to these plans, and as you can see, the plan can be confusing.

When setting up a Keogh plan, you must name a trustee. The trustee is usually a financial institution. Before attempting to establish and use your own Keogh plan, consult with an attorney who is familiar with the rules and regulations.

Pension Plans

Pension plans for your employees must be funded in good years and bad years. If you are hiring older employees, they should prefer a pension plan over a profit-sharing plan. Pension plans provide a consistent company contribution to the employee's retirement plan. Pension plans are termed as qualified plans. This means they meet the requirements of Section 401 of the Internal Revenue Code and qualify for favorable tax advantages. These tax advantages help you and your employees

PRO POINTER

If you decide to use a qualified pension plan, you must cover at least seventy-percent of your average employees. The features and benefits of these plans are extensive. For complete details on forming and using such a plan, consult with a qualified professional.

Profit Sharing

Profit-sharing plans can also be termed as qualified-plans. One advantage to you, as the employer, with a profit-sharing plan is that there is no regulation requiring you to fund the plan in bad economic years. A formula will be established to identify the amount of contributions that will be made to profit-sharing plans. The plan will also detail when contributions will be made. Many new companies prefer profit-sharing plans. Why do they like these plans? Profit-sharing plans are favored because there is no mandatory funding in years where a profit is not made.

Social Security

Did you know that Social Security benefits are taxable? Well, they can be. If an individual's adjusted gross income, tax-exempt interest, and one-half of the individual's social security benefits exceed a certain level, the Social Security benefits can be taxed. Talk to you accountant about any recent changes in Social Security benefits as they relate to taxable income.

Annuities

Annuities can be good retirement investments. These investments are safe, pay good interest rates, and the interest earned is tax-deferred until it is cashed. However, if you need access to your money early, you will have to pay a penalty for early withdrawal. If you plan to let your money work for you for seven to ten years, annuities are a safe bet.

If you decide to put your money in annuities, shop around. There are a multitude of programs open to you. If you want to investigate annuity plans for employees, talk with professionals in the field. Again, there are many options available for these programs, but there are also rules to be followed.

Some of the Considerations

Investing for your future can involve a variety of strategies. Here are some potential retirement investments: bonds, art, antiques, diamonds, gold, silver, rare coins, stocks, and mutual funds. Any of these forms of investments can return a desirable rate of return . However, many of these investments require a keen knowledge of the market. For example, if you are not an experienced coin buyer, your rare coin collection may wind up being worth little more than its face value.

For most business owners, conservative investments are the best bet for retirement. If you have some extra money you can afford to play with, you might diversify your conservative investments with some of the more exciting opportunities available. However, when you are betting on your golden years, play your cards carefully.

Some Final Words on Retirement Plans

Allow me to give you some final words on retirement plans. Retirement plans for you and your employees can be quite sophisticated. With the complexity of these plans, you should always consult experts before making decisions. Make yourself aware of your responsibilities to your employees.

There are lots of things to be considered when you are thinking about treating your employees fairly. Don't assume that part-time employees should not be entitled to some of the same benefits as your full-time employees. There probably are exceptions to part-time help, but don't make that assumption. Don't assume anything. Employees' rights and applicable laws are too important to treat lightly or make assumptions that may be wrong. Consult professionals and maintain your integrity as an employer.

C H A P T E R   N I N E T E E N

Nuts and Bolts of General Remodeling Projects

Residential remodeling is a vast field. It can range from very small jobs to massive jobs. The work involved can be basic carpentry or complex mechanical work. Once you are established, you can control the type of work that you want. This comes with time. As you earn a living and begin to making money, which are two different things, you will evolve in the eyes of your customers. Let's discuss this for a moment.

A majority of people earn a living. In my view, this means making enough money to get by and to hopefully have a few toys. It is very difficult to make "money" if you are making a living. Does this confuse you? It might. When I talk about making money, I am talking about depositing at least twice the amount of money that you need to live on. Remodelers who are good at what they do can accomplish this. Yes, you can make a small fortune, based on average-people standards, as a remodeler.

If you are a carpenter, what is your hourly wage? In my area, employed carpenters are paid, on average, $12 to $20 per hour. For Maine, this is not bad money. It wouldn't go far in New York City or LA, but it works in Maine. What does an independent contractor who is a carpenter make in the same area? There are cheap ones out there charging $17 an hour, but most serious independent carpenters are making $25-$30 per hour. Are they better off? Maybe not. You have to factor in many considerations before you can make the judgment call. Here are some of the additional expenses that independent carpenters encounter and some of their risk factors that can affect their net income:

- Supply your own truck
- Supply your own commercial truck insurance

- Supply your own liability insurance
- Supply your own tools
- Supply your own insurance for tools and supplies
- Establish your own credit for supplies
- Supply your own business license (if required in your area)
- Run the risk of code and safety violations and the resulting fees
- Learn how to do your own take-offs and estimates with no margin for error
- Develop your own legal and accounting team, even if they are independent professionals
- Design and implement your own marketing plan
- Design and implement your own advertising plan
- Set up and maintain a business address
- Establish a merchant account for accepting credit cards (this can be tough)
- Weigh the need for a professional office environment
- Hire an office administrator or run you own office
- Establish a computerized job-costing and estimating system (nearly essential in today's world)
- Hire a payroll company for employee payments, work alone, or do your own payroll
- Deal with Worker's Compensation regulations if you hire anyone as an employee
- Consider self-employed benefit packages and similar packages for employees
- Do your own billing for work done or hire someone to do it
- Accept the fact that some customers will not pay their bills
- Be prepared for potential lawsuits from your work or that of your subcontractors
- Purchase, equip, and maintain company trucks as needed for expansion
- Maintain inventory control for tax purposes and to prevent theft by employees
- Maintain accurate tax records for your accountant
- Decide on the most favorable business structure, such as a sole proprietorship, "S" corporation, "C' corporation, or partnership.

- Advertising expenses, such as phone directories
- Tool and equipment expenses
- Storage facilities for materials and tools
- Are you getting the picture?

You could review the list above and run away from going into business for yourself. Running is rarely the answer, but you might not yet be ready. If you have not considered the points above, you should. Being in business for yourself is now always what people perceive it to be.

I roll out of bed by 6:00 a.m and begin my work. It is not uncommon for me to still be seeing a clock at 1:00 a.m. before I go to bed. These are a lot of hours, but I am used to it. Don't get me wrong, you don't have to be as aggressive as I am. But you will have field work during the day and phone calls with customers and subs at night. This will not be a walk in the park if you want to make more than a living.

If you are only making $10 per hour more by being in business for yourself, the financial portion of your decision is a mistake. However, there is the element of freedom and control that you gain by being self-employed that cannot have a fixed price put on it. And, if you learn to build a business and have people working effectively for you, the rewards can be quite rich.

As you move into becoming a remodeling contractor, you will encounter all sorts of work. The purpose of this chapter is to discuss, in brief, the various opportunities in front of you. With this said, let's look at some general remodeling topics and evaluate my experiences with each type so that you may be better prepared to make wise decisions for yourself.

Decks

PRO POINTER

A deck is a fast job to do with limited expense and a quick payday from the customer.

Decks are an excellent source of quick money. Is this remodeling? Not really. It is more of a home improvement, but remodeling contractors are often called to build decks. Most decks require only moderate carpentry skills. Some companies specialize in decks, as I specialize in bathroom and kitchens. You may have trouble competing with high-volume deck builders, but it can be done. I did it in Virginia. The key is knowing how to sell the job.

If you don't have sales skills, you should strive to gain them as quickly as possible if you are going into business for yourself. This applies to all aspects of your new business, not just decks. You can sell on the fear factor, the experience factor, the quality factor, and so forth. The decision of which method to use will depend on you and your skills. But, if you can't sell, you can't work as an independent remodeling contractor.

Risks are minimal with decks. It is usually all good if you have a clue about construction. For any decent carpenter and most general contractors, decks are a good card to deal into your hand.

Gazebos and Porches

Gazebos and porches are not high-volume deals, but they are good jobs to take. In some areas, screened porches are a high-volume opportunity. These are simple structures to build. They are not true remodeling. Like decks, they are home improvements, but they are cash cows for savvy contractors. This area of expertise is well worth getting into.

Small Jobs

Small jobs, such as installing gutters or replacing a screened door, are often shrugged off by contractors. This is a mistake. Small jobs often turn into large jobs. One of my means of success if using my plumbing and electrical divisions to get full remodeling jobs. My people may respond to a leaking faucet and leave with a bathroom remodel. How do they do this? They are trained to do so.

PRO POINTER

If customers like your company, they are more likely to hire the company to do larger jobs.

It is tempting to blow off service calls and tiny jobs, but you shouldn't. Every house you can put people in is an opportunity for future work. Set up a single day a week where you do small jobs and then use the small jobs as an opening for larger jobs. If you have to, set the jobs up as "Will-Call Jobs". This is where you can't schedule a date or time, but you will call the customer if you have an early afternoon or a busted day due to bad deliveries. You will lose some of the jobs, but you will get many of them. The key is getting into the home and in front of the homeowner. If you do this with modest sales skills, you can grow your business quickly.

Room Additions

Room additions are a gray job in most cases. These can be worth tens of thousands of dollars, and the work is clean and simple, most of the time. You will probably need some subcontractors. At the least, you are likely to need a site-work contractor, someone to dig footings, and someone to pour the foundation. Depending on your business structure, you may need the following subs:

- Heating mechanics
- Plumbers
- Electricians
- Drywall contractors
- Painters
- Flooring contractors

With my company, we do much of this type of work with in-house people. But, we do sub out the site work, footings, and foundation work. The electrical work is subbed out, and the heating work is sometimes subbed out. In any event, room additions are a tremendous opportunity for any home-improvement contractor or remodeling contractor.

Attic Conversions

PRO POINTER

Don't go into attic conversions unless you have extensive experience in the field.

Attic conversions can be both profitable and challenging. You may have to remove a roof and replace it. A dormer may need to be added. Finding a suitable location for stairs can pose problems. Reinforcing ceiling joists can lower headroom and increase costs. This type of work takes special expertise.

I have done a lot of attics, but they are a difficult job to do well. You have to protect the ceilings below you, the roof above you, and provide adequate lighting, ventilation, mechanical services, and so forth. This is not a good area for a rookie to roam in. However, it is a profitable venture for experienced people.

Basement Conversions

Basement conversions are rarely a good investment for homeowners, but they tend to be a safe deal for remodelers. Wet basements are the biggest problem for remodelers, but they can be overcome with reasonable expense. Many homeowners will want you to remodel their basements into configurations that go against code requirements. For example, they may want a bedroom in a buried basement. This is a problem. Without proper egress, such as windows or doors, a bedroom is rarely allowed. You will have to decide where you will stand in this type of situation. You can call the room a study and get away with it, but can you sleep with yourself when someone dies in a fire due to your actions? I can't answer this question. Only you can determine your ethics.

There is decent money in basement conversions, but you may need some special skills to deal with moisture problems. This is not hard to gain if you read enough material on the subject. Generally speaking, basement conversions are not difficult to do.

Garages

Garages are a common request of contractors. Many contractors sub out the site work and the slab. Do you have these subcontractors? If not, they might be worth finding. However, garages are a very competitive field. A typical garage has dimensions of 24' x 24'. This is easy for a homeowner to compare prices on.

We don't do a lot of garages. My reason for this decision is my quality requirements versus the sales mechanisms of other contractors. If you get established as a garage contractor, you can make good money off of them quickly. My focus is on more technical work, but there is nothing wrong with doing garages if you are geared up for them. This phase of work is certainly one well worth considering.

Basic Remodeling

Basic remodeling can include almost anything. What will you do? This is a question for you to ask yourself. The long story made short is that there is more hourly profit in remodeling than there is in new construction. There are multiple reasons for this. I often say that any good remodeler can build a house, but not any good builder can remodel a home. This is due to the challenges that remodelers face. Keep this in mind when hiring your people.

Whatever form of remodeling you engage in, the opportunity is there for you. If you establish yourself as a specialist in some sort of field, you will have an added advantage. Even if you are a cookie-cutter remodeler, you can do very well and make money, rather than just making a living. It does not take large crews to make major money. But, it does take the right people. Choose your crews carefully and enjoy the financial gains that are there for professional remodelers.

PRO POINTER

Remodeling requires a lot more thought and experience than building a new house.

C H A P T E R T W E N T Y

Jobsite Safety

Construction is one of the most dangerous industries in the nation, vying with mining and meatpacking as workplaces where the most fatalities occur each year. The year 2001 marked a high point in construction-related jobsite deaths, reaching 1225. You or your workers do not want to be part of those statistics.

A poor safety record affects your company in a number of ways. Ignoring basic safety practices can cause unnecessary human suffering and pain. Secondarily, the premiums you pay to the state for Worker's Compensation insurance will increase your overhead costs quite a bit (and once your rates go up, it will take you three years of good safety records to get a reduction in these costs). And thirdly–something contractors often fail to realize–if one of your key workers or supervisors is injured and out of work, your productivity will take a nosedive until such time as you can temporarily fill the gap. Lastly, if an Occupational Safety and Health Administration (OSHA) inspector comes to your job site and cites you for a serious violation, your fine can be as much as $70,000, enough to drive a small company into bankruptcy. So safety really does pay off.

Recognizing where and how most accidents on the job occur is the first step in prevention. Based upon statistics furnished by the Department of Labor's Occupational Safety and Health Administration, the most frequently reported accidents were as follows:

- Falls from elevated areas–from scaffolding or from working at heights without some kind of fall protection.

- Being struck by an object or machine. This is why equipment is required to have that beeping device to sound an alarm and alert nearby workers.
- Being caught in between. Many injuries occur when a piece of equipment crushes a worker against a building or material stockpile because neither operator nor worker is paying attention to the activity around them.
- Electrical hazards caused by nonexistent or poorly grounded electrical cables, excavating equipment hitting power lines, or exposed or poorly insulated wiring.

Let's Talk About OSHA

OSHA regularly conducts construction site visits, often in conjunction with their state counterparts, since most states have followed suit with the federal government and formed their own OSHA agencies. OSHA inspectors are required by law to perform routine inspections on construction sites and to conduct an inspection in response to a complaint about a builder who is operating without regard for safety on a project. OSHA inspectors are required by law to issue citations to builders for standard violations, and several of these violations carry with them monetary penalties.

You as a builder will be subjected to the rules and regulations of both federal and state OSHA organizations. You can become familiar with the rules and regulations by obtaining a copy of a booklet known as 29CFR 1926, Subpart L by calling your local U.S. Department of Labor office. OSHA also has a website that contains a lot of good safety pointers.

PRO POINTER

Some of OSHA's most important safety regulations deal with trenching operations and personal protective equipment (hard hats, eye and ear protection, and so forth).

Trenching

A high-profile case involving a trenching accident was picked up by the newspaper and TV stations in January 2004. It involved the death of a plumber working in an unprotected ditch. OSHA has strict rules about trenching, and depending upon the depth of the trench and the type of soil encountered, you either need a trench box to prevent the walls from caving in, or you have to slope the excavation sufficiently to prevent any cave-ins. In the case of this plumber, he was working in a deep trench installing a sanitary pipe and noticed water seeping through the walls, but before he could yell for help, he was buried under 8 feet of dirt and died before help arrived. His family was thinking about suing the plumbing contractor and possibly filing criminal charges against the owner. OSHA is considering a huge fine—so it looks like this company will go out of business and the owner might possibly go to jail.

This may be an extreme case, but if certain precautions aren't taken during excavation and deep trenching operations, it could happen to you. Go to the OSHA website and get familiar with the correct way to trench for foundations or underground utilities-it could save your life or the life of one of your workers.

Personal Protection

Hard hats, goggles, and earplugs are all part of OSHA's personal protection concerns, but it doesn't stop there. According to the federal standards, only steel-tipped, thick-soled work shoes are allowed on the construction site. Anyone wearing sneakers or running shoes is not permitted to work. And you know those polyester shirts that are so easy to wash and usually require no ironing? They are banned when anyone is welding or burning with an open torch. If these polyester shirts catch fire, they continue to smolder and burn even as the flames are being extinguished. Cotton shirts, on the other hand, do not smolder and any fire is quickly extinguished.

Boom boxes and 200-watt car speakers reduce hearing, and so does operating some noisy construction equipment, jackhammers, and the like. Get some inexpensive earplugs for your workers.

And what about eye protection? How often have you seen masons cutting bricks, rebars, or blocks with their carbon-tipped power saws without any eye protection? Usually the excuse for not wearing protective eyewear is, "Well, I'm only going to cut one block", but that may just be the one that blinds them. Don't chance losing your eyesight—wear safety glasses when the occasion arises and require your workers to do the same.

PRO POINTER

Each year about 100,000 eye injuries are caused by workers who fail to wear proper eye wear.

Developing a Good Safety Program

First of all you as the owner must be aware of good safety practices before you can instruct others. If you need help, there are plenty of books on construction safety practices, and your local OSHA office will also be able to provide you with some material or point the way to other sources. There are companies that specialize in preparing and administering safety programs but they are expensive and might not be suited to your operation.

You can write your own safety program, listing the "What to Do's" and the "What Not to Do's" and giving it to each employee. Be prepared to answer questions, because there will probably be a lot of them. When you establish rules for wearing work shoes, safety glasses, earplugs, and other personal protective equipment, enforcement is what will make or break your safety program. If you pass someone cutting a brick or block with a masonry saw and they are not wearing safety glasses, stop them immediately and tell them to get the proper glasses or stop cutting. When someone is observed wearing sneakers, tell him to get the proper shoes-otherwise he can't work on the site. This is a tough one if you really need that worker to finish a particular operation that day, but unless you tell him to change into a pair of acceptable shoes or leave the job, the rest of the crew will know you are not serious about safety.

PRO POINTER

Everyone wants to work safely, and as the owner of the business you have the power to provide each employee with a safe environment in which to work.

Don't let a safety violation continue. The next minute a serious accident or a fatality could occur, and for many years afterward you will live with that decision on your conscience.

Index

Entries in italics indicate figures

ABOUT THE AUTHOR

R. Dodge Woodson knows the home construction business inside and out. The owner of The Masters Group, Inc., of Brunswick, Maine, he has been a successful remodeling contractor for almost a quarter of a century. He is also a licensed master plumber, a Class A builder who has constructed over 60 homes in a single year, and a Designated Broker, the highest classification of professional real estate licensure available. With nearly 30 years of experience in construction and remodeling, Woodson is a true business insider whose advice has been tried and tested in the field.